Springer Series in

MATERIALS SCIENCE 102

Springer Series in

# MATERIALS SCIENCE



The Springer Series in Materials Science covers the complete spectrum of materials physics, including fundamental principles, physical properties, materials theory and design. Recognizing the increasing importance of materials science in future device technologies, the book titles in this series reflect the state-of-the-art in understanding and controlling the structure and properties of all important classes of materials.

88 **Introduction to Wave Scattering, Localization and Mesoscopic Phenomena**
By P. Sheng

89 **Magneto-Science**
Magnetic Field Effects on Materials: Fundamentals and Applications
Editors: M. Yamaguchi and Y. Tanimoto

90 **Internal Friction in Metallic Materials**
A Handbook
By M.S. Blanter, I.S. Golovin, H. Neuhäuser, and H.-R. Sinning

91 **Time-dependent Mechanical Properties of Solid Bodies**
By W. Gräfe

92 **Solder Joint Technology**
Materials, Properties, and Reliability
By K.-N. Tu

93 **Materials for Tomorrow**
Theory, Experiments and Modelling
Editors: S. Gemming, M. Schreiber and J.-B. Suck

94 **Magnetic Nanostructures**
Editors: B. Aktas, L. Tagirov, and F. Mikailov

95 **Nanocrystals and Their Mesoscopic Organization**
By C.N.R. Rao, P.J. Thomas and G.U. Kulkarni

96 **GaN Electronics**
By R. Quay

97 **Multifunctional Barriers for Flexible Structure**
Textile, Leather and Paper
Editors: S. Duquesne, C. Magniez, and G. Camino

98 **Physics of Negative Refraction and Negative Index Materials**
Optical and Electronic Aspects and Diversified Approaches
Editors: C.M. Krowne and Y. Zhang

99 **Self-Organized Morphology in Nanostructured Materials**
Editors: K. Al-Shamery, S.C. Müller, and J. Parisi

100 **Self Healing Materials**
An Alternative Approach to 20 Centuries of Materials Science
Editor: S. van der Zwaag

101 **New Organic Nanostructures for Next Generation Devices**
Editors: K. Al-Shamery, H.-G. Rubahn, and H. Sitter

102 **Photonic Crystal Fibers**
Properties and Applications
By F. Poli, A. Cucinotta, and S. Selleri

103 **Polarons in Advanced Materials**
Editor: A.S. Alexandrov

Volumes 40–87 are listed at the end of the book.

F. Poli A. Cucinotta S. Selleri

# Photonic Crystal Fibers

## Properties and Applications

With 129 Figures

Springer

Federica Poli
Dipartimento di Ingegneria dell'Informazione
Universita degli Studi di Parma
Viale G.P. Usberti 181/A - Campus Universitario
I-43100 Parma, Italy

Annamaria Cucinotta
Dipartimento di Ingegneria dell'Informazione
Universita degli Studi di Parma
Viale G.P. Usberti 181/A - Campus Universitario
I-43100 Parma, Italy

Stefano Selleri
Dipartimento di Ingegneria dell'Informazione
Universita degli Studi di Parma
Viale G.P. Usberti 181/A - Campus Universitario
I-43100 Parma, Italy

*Series Editors:*

Professor Robert Hull
University of Virginia
Dept. of Materials Science and Engineering
Thornton Hall
Charlottesville, VA 22903-2442, USA

Professor Jürgen Parisi
Universität Oldenburg, Fachbereich Physik
Abt. Energie- und Halbleiterforschung
Carl-von-Ossietzky-Strasse 9–11
26129 Oldenburg, Germany

Professor R. M. Osgood, Jr.
Microelectronics Science Laboratory
Department of Electrical Engineering
Columbia University
Seeley W. Mudd Building
New York, NY 10027, USA

Professor Hans Warlimont
Institut für Festkörper-
und Werkstofforschung,
Helmholtzstrasse 20
01069 Dresden, Germany

A C.I.P. Catalogue record for this book is available from the Library of Congress

ISSN 0933-033x

ISBN 978-90-481-7609-0
ISBN 978-1-4020-6326-8 (e-book)

Published by Springer,
P.O. Box 17, 3300 AA Dordrecht, The Netherlands

www.springer.com

# Contents

# Preface

Photonic crystal fibers, also known as microstructured or holey fibers, have recently generated great interest in the scientific community thanks to the new ways provided to control and guide light, not obtainable with conventional optical fibers. Proposed for the first time in early 90's, photonic crystal fibers have driven an exciting and irrepressible research activity all over the world, starting in the telecommunication field and then touching metrology, spectroscopy, microscopy, astronomy, micromachining, biology and sensing.

A variety of very interesting publications and high level books have been already presented, describing the different kinds of these new fibers, the physics of their behavior, as well as a huge range of design tools. These aspects will not be considered again in this work.

This book, instead, is intended to provide an expert guidance through the properties of photonic crystal fibers, with a specific focus on the telecommunication aspects. Although standard fibers for telecommunication can rely on a well-established technology and standard fiber based devices and systems represent a consolidated reality, hardly replaceable, the authors believe that photonic crystal fibers can revolutionize the field of guided optics and its applications, even if much easier and close opportunities can be foreseen in many other fields. This belief gets firmer when considering signal processing and specific functions rather than the usage of photonic crystal fibers in long distance transmission.

The long expertise of the authors in fiber based device analysis is reflected in a deep analysis aimed to practically understand how the physical and geometrical characteristics of these new fibers can be tailored to achieve the goal of ad hoc performances. The study, mainly performed with the help of the finite element method, a powerful numerical approach the authors are very expert in, has enabled to understand how best to optimize the fiber design, always keeping in mind actual possibilities and limits of photonic crystal fiber fabrication technology.

This book will thus benefit researchers approaching this very dynamic and evolving subject with the interest to explore this field of telecommunication, looking at current as well as emerging applications.

# Acknowledgements

The authors would like to thank all the people who have actively participated in their research activity regarding photonic crystal fibers in the last years. A special thanks is due to Matteo Foroni for his constant support in the experimental activity and in the theoretical analysis, and for his fundamental help in the book writing; to Lorenzo Rosa for his valuable work on the finite element code improvement and development; to Luca Vincetti for the fruitful and stimulating discussions.

The authors are grateful to Crystal Fiber A/S for providing all the pictures of the photonic crystal fibers inserted in this book.

# Introduction

Until recently, an optical fiber was a solid thread surrounded by another material with a lower refractive index. Today, photonic crystal fibers (PCFs) are established as an alternative fiber technology. PCFs, which have been first demonstrated in 1995, are optical fibers with a periodic arrangement of low-index material in a background with higher refractive index. The background material in PCFs is usually undoped silica and the low-index region is typically provided by air-holes running along their entire length.

Two main categories of PCFs exist: high-index guiding fibers and photonic bandgap ones.

PCFs belonging to the first category are more similar to conventional optical fibers, because light is confined in a solid core by exploiting the modified total internal reflection mechanism. In fact, there is a positive refractive index difference between the core region and the photonic crystal cladding, where the air-hole presence causes a lower average refractive index. The guiding mechanism is defined as "modified" because the cladding refractive index is not a constant value, as in standard optical fibers, but it changes significantly with the wavelength.

This characteristic, as well as the high refractive index contrast between silica and air, provides a range of new interesting features. Moreover, a high design flexibility is one of the distinctive properties of PCFs. In particular, by changing the geometric characteristics of the air-holes in the fiber cross-section, that is, their dimension or position, it is possible to obtain PCFs with diametrically opposite properties. For example, PCFs with a small silica core and large air-holes, that is, with a high air-filling fraction in the transverse section, have better nonlinear properties compared with conventional optical fibers, and so they can be successfully used in many applications, like supercontinuum generation. On the contrary, fibers can be designed with small air-holes and large hole-to-hole distances, in order to obtain a large modal area, useful for high-power delivery. Differently from standard fibers, PCFs with proper geometric characteristics can be endlessly single mode, that is, only the fundamental mode is guided, regardless of the wavelength. In addition, a significant asymmetry can be introduced in a simple way in the PCF core, thus creating fibers with very high level of birefringence. Moreover, the PCF dispersion properties can be tailored with high flexibility, that is, it is possible to move the zero-dispersion wavelength to the visible range, as well as to obtain dispersion curves ultraflattened or with a strong negative slope.

When the PCF core region has a lower refractive index than the surrounding photonic crystal cladding, light is guided by a mechanism different from total internal reflection, that is, by exploiting the presence of the photonic bandgap (PBG). In fact, the air-hole microstructure which constitutes the PCF cladding is a two-dimensional photonic crystal, that is a material with periodic dielectric properties characterized by a photonic bandgap, where light in certain wavelength ranges cannot propagate. The PBG effect can be also found in nature, since it is responsible, for example, of the beautiful and bright colors seen in butterfly wings. PCFs with a low index core are created by introducing a defect in the photonic crystal structure, for example, an extra air-hole or an enlarged one, and light is confined because the PBG makes propagation in the microstructured cladding region impossible. This guiding mechanism cannot be obtained in conventional optical fibers and it opens a whole new set of interesting possibilities.

In particular, light can be guided in air in PCFs with a hollow core, thus providing numerous promising applications, such as low-loss guidance and high-power delivery, without the risk of fiber damage. Moreover, air-guiding PCFs are almost insensitive to bending, even for small bending diameter values, and they present extreme dispersion properties, highly dominated by the waveguide component. Finally, when filled with proper gases or liquids, hollow-core PCFs can be successfully employed in sensor applications or for nonlinear optics.

Since their first demonstration, PCFs have been the object of an intense research activity by the most important groups all around the world. In fact, it is particularly intriguing to study the new light-guiding mechanisms offered by PCFs and the innovative properties related to the presence of the PBG. Moreover, the possibility of modifying the air-hole geometry in the fiber cross-section is limited only by the technological feasibility of the designed PCFs. It is also very interesting to investigate how the PCF properties can be influenced by the changes of the geometric characteristics and "how far" it is possible to go from the well-known and established properties of standard optical fibers.

The research activity described in this book is set in this context, which is in continuous evolution and characterized by a great scientific excitement. The aim of the research carried out and here reported has been to accurately study, and thus to deeply understand, the light-guiding mechanisms exploited in this new kind of optical fibers. PCFs with unusual guiding, dispersion, and amplification properties have been designed by exploring different air-hole arrangements in the fiber cross-section. This has been done with a constant attention to the possible applications of the proposed PCFs, in the field of the optical communications. Moreover, the performances of the traditional optical fibers have been always considered as a useful comparison parameter, in order to evaluate the effective advantages offered by these new fibers. Finally, the results of these studies have been presented in a critical way, that is, by underlining the possible drawbacks, which are usually related to the PCF attenuation, which is still higher than that of the conventional optical fibers.

The book is organized in six chapters. Chapter 1 is a general presentation of the PCF innovative characteristics. Starting from the description of the properties of photonic crystals, materials with a refractive index periodic distribution, the passage from conventional optical fibers to photonic crystal ones is explained. After describing the two light-guiding mechanisms exploited in PCFs, the advantages offered by this new fiber type with respect to the conventional ones are discussed. Then, some meaningful examples of PCFs with unusual guiding, dispersion, and nonlinear properties, proposed in the literature and successfully used in many applications, are reported. Moreover, the different loss mechanisms are presented for both solid- and hollow-core PCFs, since attenuation is still the main drawback which affects this new kind of optical fibers. Once a significant loss reduction is obtained, which can be reached by improving the fabrication process described in the final part of Chapter 1, these new fibers will enter in the market in a significant way.

In Chapters 2–6 the main results of the research activity carried out by the authors in the past years are presented. In each chapter, results concerning the same topic, that is, guiding, dispersion, or amplification properties, are collected. It is important to underline that all the analyses reported in this book have been developed by mainly using the finite element method (FEM), in particular, a full-vector modal solver, as described in Appendix A. This numerical method is particularly suitable to study PCFs, since fibers with any refractive index profile, as well as any air-hole arrangement in the transverse section, including the nonperiodic ones, can be analyzed.

Chapter 2 summarizes the results concerning the PCF guiding properties, which directly come from the complex propagation constant of the guided modes. First of all, the study of the influence of the geometric parameters on the fundamental guided-mode characteristics in a new kind of PCF, with a square lattice of air-holes, is reported. Moreover, the modal cutoff analysis of these PCFs is presented. The same method has been successfully applied to study the single-mode regime of a new kind of triangular PCFs, which have a wide silica core and a large modal area. In fact, it is important to investigate the trade-off between the effective area and the cutoff of the fundamental guided mode, in order to successfully exploit these large-mode area fibers in practical applications. In the final part of the chapter the study of the guiding, leakage, and birefringence properties of hollow-core PCFs with a modified honeycomb lattice, which guide light by exploiting the PBG effect, is reported. Air-guiding has been demonstrated in fibers with a larger bandgap with respect to that obtained with triangular lattices.

The design of PCFs with innovative dispersion properties is described in Chapter 3. In fact, it is possible to significantly change the waveguide contribution to the dispersion parameter by properly changing the geometric characteristics of the air-holes in the cross-section. Triangular PCFs characterized by a high air-filling fraction, that is, with large air-holes and small hole-to-hole spacing, have been designed to compensate the anomalous dispersion and the dispersion slope of single-mode fibers around 1550 nm, as it is reported at the beginning of the chapter. Then, the dispersion properties of fibers with a square lattice of air-holes, obtained with different values of the geometric parameters, are discussed and compared with those of triangular PCFs. In the second part of the chapter the design of triangular PCFs with completely different dispersion characteristics, that is with flattened dispersion curve and zero-dispersion wavelength around 1550 nm, which can be exploited for nonlinear applications, is described. The cross-section geometry around the core of the triangular PCFs has been modified in two different ways, in order to obtain the desired dispersion properties and a small effective area, that is a high nonlinear coefficient.

Chapter 4 deals with the PCF nonlinear properties. Firstly, supercontinuum generation is described, since it is one of the most interesting applications of highly nonlinear fibers. The most important results, both experimental and theoretical, presented so far in the literature have been collected to explain the characteristics of this complex combination of nonlinear phenomena. Then, the possibility to exploit PCFs for optical parametric amplification, which is based on the highly efficient nonlinear effect of four-wave mixing, has been investigated. A high fiber nonlinearity as well as a low dispersion slope are fundamental aspects for a successful design of an optical parametric amplifier, that is with a high and broadband gain. PCFs are suitable for this kind of amplification, since they offer the possibility to engineer the dispersion curve and to obtain enhanced nonlinear properties. In the final part of the chapter a different kind of PCF, that is the one with hollow core, has been considered, even if these fibers present negligible nonlinear characteristics. The nonlinear coefficient of hollow-core fibers with modified honeycomb lattice has been evaluated, showing that also the nonlinear contribution of air should be taken into account.

An important part of the developed research activity concerns the possibility of using PCFs for Raman amplification, which has become more and more relevant in the past years for optical communication systems. Chapter 5 is completely devoted to this topic. In particular, two meaningful parameters, that is the Raman effective area and the Raman gain coefficient, have been introduced to describe the PCF Raman performances. All-silica PCFs as well as germania-doped ones have been considered, in order to design nonlinear fibers with enhanced performances for Raman amplification. Moreover, the Raman properties of tellurite-based triangular fibers and of honeycomb PCFs with a germania-doped solid core which guide light by means of the PBG have been considered and discussed. A complete model of PCF-based Raman amplifiers, proposed to study the Raman amplification process in a PCF is fully described in the second part of the chapter. The gain and noise performances of different triangular PCF Raman amplifiers have been analyzed, in order to underline the influence of the geometric parameters which characterize the fiber cross-section. Moreover, the performances of Raman amplifiers based on triangular PCFs have been investigated, by evaluating the potential improvements obtainable with a reduction of the background losses. A further study, described in the last part of the chapter, has been performed on Raman amplifiers based on low-loss triangular PCFs when multiple pumps are used. Different pump wavelengths and power distributions have been considered, with the aim to reduce as much as possible the gain spectrum ripple.

In Chapter 6 the amplification in erbium-doped PCFs is discussed. In fact, active fibers with superior characteristics with respect to standard ones can be obtained by a proper PCF design. The amplification properties have been studied with a numerical model which combines the full-vector modal solver with a population and propagation rate equation solver. In particular, in the analysis presented here the model has been applied to design erbium-doped triangular PCFs which exhibit high gain values and low losses when spliced with a standard single-mode fiber.

The study presented here does not pretend to be exhaustive of all the possible telecommunications and PCF applications which, for sure, will further increase and improve in the future. The intent, instead, is to collect and provide examples, based on the authors' experience of the potentialities and the limits of PCF exploitation, which can hopefully lead to actual and practical designs of new optical devices.

# Chapter 1

# Basics of photonic crystal fibers

In this chapter, starting from the description of the characteristics of photonic crystals, materials with a refractive index periodic distribution, the passage from conventional optical fibers to photonic crystal ones, introduced for the first time in 1995, is explained.

Then, the two light-guiding mechanisms are presented. In solid-core photonic crystal fibers, where light is confined in a higher refractive index region, modified total internal reflection is exploited, which is quite similar to the guiding mechanism of standard optical fibers. Instead, when the light is confined in a region with a refractive index lower than that of the surrounding area, as in hollow-core fibers, it is necessary the presence of the photonic bandgap (PBG).

One of the most important advantages offered by photonic crystal fibers (PCFs) is the high design flexibility. In fact, by changing the geometric characteristics of the fiber cross-section, such as the air-hole dimension or disposition, it is possible to obtain fibers with diametrically opposed optical properties. PCFs with unusual guiding, dispersion, and nonlinear properties can be designed and successfully used in various applications, as reported in this chapter.

The main drawback which affects this new kind of fibers is related to the attenuation, which is higher than that of conventional optical fibers. The different loss mechanisms are thus analyzed for both solid- and hollow-core photonic crystal fibers.

In general, a loss reduction for PCFs can be obtained by improving the fabrication process, reported in the last part of the chapter. The stack-and-draw process is described with other fabrication techniques, like extrusion, usually employed to realize fibers with materials different from silica, such as soft-glasses or polymers. Once reached the technological maturity, the advantages offered by PCFs with respect to conventional fibers will be completely exploited for different applications, as described in the final part of the chapter, and the new fibers will enter concretely in the market.

## 1.1 From conventional optical fibers to PCFs

Optical fibers, which transmit information in the form of short optical pulses over long distances at exceptionally high speeds, are one of the major technological successes of the 20th century. This technology has developed at an incredible rate, from the first low-loss single-mode waveguides in 1970 to being key components of the sophisticated global telecommunication network. Optical fibers have also non-telecom applications, for example, in beam delivery for medicine, machining and diagnostics, sensing, and a lot of other fields. Modern optical fibers represent a careful trade-off between optical losses, optical nonlinearity, group velocity dispersion, and polarization effects. After 30 years of intensive research, incremental steps have refined the capabilities of the system and the fabrication technology nearly as far as they can go.

The interest of researchers and engineers in several laboratories, since the 1980s, has been attracted by the ability to structure materials on the scale of the optical wavelength, a fraction of micrometers or less, in order to develop new optical medium, known as *photonic crystals*. Photonic crystals rely on a regular morphological microstructure, incorporated into the material, which radically alters its optical properties [1.1]. They represent the extension of the results obtained for semiconductors into optics. In fact, the band structure of semiconductors is the outcome of the interactions between electrons and the periodic variations in potential created by the crystal lattice. By solving the Schrödinger's wave equation for a periodic potential, electron energy states separated by forbidden bands are obtained. PBGs can be obtained in photonic crystals, where periodic variations in dielectric constant, that is in refractive index, substitute variations in electric potential, as well as the classical wave equation for the magnetic field replaces the Schrödinger's equation [1.2].

PBG, originally predicted in 1987 by Sajeev John, from University of Toronto, and Eli Yablonovitch, from Bell Communications Research, has become the really hot topic in optics in the early 1990s. The idea was to build the right structures, in order to selectively block the transmission of photons with energy levels, that is wavelengths, corresponding to the PBGs, while allowing other wavelengths to pass freely. Moreover, slight variations in the refractive index periodicity would introduce new energy levels within the PBG, as it happens with the creation of energy levels within the bandgap of conventional semiconductors.

Unfortunately, building the right structures has proved extremely difficult. The first PBG material was created in 1991 by Yablonovitch and his colleagues by drilling holes with a diameter of 1 mm in a block of material with a refractive index of 3.6. Since the bandgap wavelength is of the order of the spacing between the air-holes in the photonic crystal, this structure had a bandgap in the microwave region.

In 1991, Philip Russell, who was interested in Yablonovitch's research, got his big "crazy" idea for "something different," during CLEO/QELS conference [1.2]. Russell's idea was that light could be trapped inside a fiber hollow core by creating a two-dimensional photonic crystal in the cladding, that is a periodic wavelength-scale lattice of microscopic air-holes in the glass. The basic principle is the same which is the origin of the color in butterfly wings and peacock feathers, that is all wavelength-scale periodic structures exhibit ranges of angle and color, *stop bands*, where incident light is strongly reflected. When properly designed, the photonic crystal cladding running along the entire fiber length can prevent the escape of light from the hollow core. These new fibers are called PCFs, since they rely on the unusual properties of photonic crystals.

The first fiber with a photonic crystal structure was reported by Russell and his colleagues in 1995 [1.3]. Even if it was a very interesting research development, the first PCF did not have a hollow core, as shown in Fig. 1.1, and, consequently, it did not rely on a photonic bandgap for optical confinement. In fact, in 1995 Russell's group could produce fiber with the necessary air-hole triangular lattice, but the air-holes were too small to achieve a large air-filling fraction, which is fundamental to realize a PBG. Measurements have shown that this solid-core fiber formed a single-mode waveguide, that is only the fundamental mode was transmitted, over a wide wavelength range. Moreover, the first PCF had very low intrinsic losses, due to the absence of doping elements in the core, and a silica core with an area about ten times larger

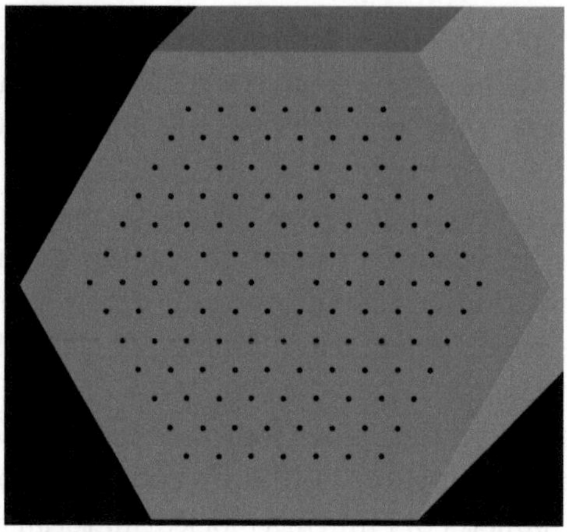

Figure **1.1**: Schematic of the cross-section of the first solid-core photonic crystal fiber, with air-hole diameter of 300 nm and hole-to-hole spacing of 2.3 μm, proposed in [1.3].

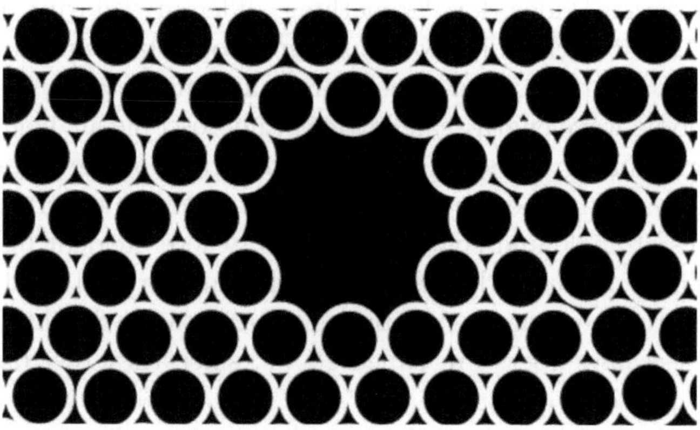

Figure **1.2**: Schematic of the cross-section of the first hollow-core PCF, with hole-to-hole spacing of 4.9 μm and core diameter of 14.8 μm, proposed in [1.4].

than that of a conventional single-mode fiber (SMF), thus permitting a corresponding increase in optical power levels.

After moving his research group to the University of Bath in 1996, where PCF fabrication techniques were steadily refined, Russell and his co-workers were able to report, in 1999, the first single-mode hollow-core fiber, in which confinement was due by a full two-dimensional PBG, as reported in Fig. 1.2.

They realized that the photonic bandgap guiding mechanism is very robust, since light remains well confined in the hollow core, even if tight bends are formed in the fiber. However, it is highly sensitive to small fluctuations in the fiber geometry, for example, to variations in the air-hole size.

Initial production techniques were directed simply at the task of making relatively short lengths of fiber in order to do the basic science, but many research teams are now working hard to optimize their PCF production techniques, in order to increase the lengths and to reduce the losses.

## 1.2 Guiding mechanism

In order to form a guided mode in an optical fiber, it is necessary to introduce light into the core with a value of $\beta$, that is the component of the propagation constant along the fiber axis, which cannot propagate in the cladding. The highest $\beta$ value that can exist in an infinite homogeneous medium with refractive index $n$ is $\beta = nk_0$, being $k_0$ the free-space propagation constant. All the smaller values of $\beta$ are allowed. A two-dimensional photonic crystal, like any other material, is characterized by a maximum value of $\beta$ which can propagate. At a particular wavelength, this corresponds to the fundamental mode of an infinite slab of the material, and this $\beta$ value defines the effective refractive index of the material.

### 1.2.1 Modified total internal reflection

It is possible to use a two-dimensional photonic crystal as a fiber cladding, by choosing a core material with a higher refractive index than the cladding effective index. An example of this kind of structures is the PCF with a silica solid core surrounded by a photonic crystal cladding with a triangular lattice of air-holes, shown in Fig. 1.3. These fibers, also known as index-guiding PCFs, guide light through a form of total internal reflection (TIR), called modified TIR. However, they have many different properties with respect to conventional optical fibers.

#### Endlessly single-mode property

As already stated, the first solid-core PCF, shown in Fig. 1.1, which consisted of a triangular lattice of air-holes with a diameter $d$ of about 300 nm and a hole-to-hole spacing $\Lambda$ of 2.3 μm, did not ever seem to become multi-mode in

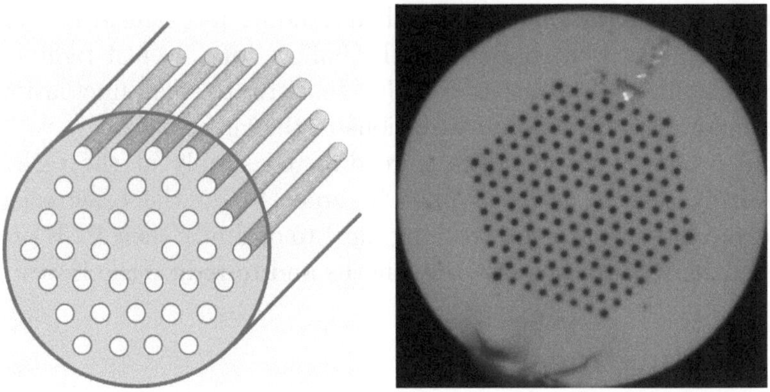

(a) (b)

Figure **1.3**: (a) Schematic of a solid-core PCF with a triangular lattice of air-holes, which guides light for modified total internal reflection. (b) Microscope picture of a fabricated solid-core triangular PCF, kindly provided by Crystal Fiber A/S.

the experiments, even for short wavelengths. In fact, the guided mode always had a single strong central lobe filling the core [1.5].

Russell has explained that this particular endlessly single-mode behavior can be understood by viewing the air-hole lattice as a modal filter or "sieve" [1.5]. Since light is evanescent in air, the air-holes act like strong barriers, so they are the "wire mesh" of the sieve. The field of the fundamental mode, which fits into the silica core with a single lobe of diameter between zeros slightly equal to $2\Lambda$, is the "grain of rice" which cannot escape through the wire mesh, being the silica gaps between the air-holes belonging to the first ring around the core too narrow. On the contrary, the lobe dimensions for the higher-order modes are smaller, so they can slip between the gaps. When the ratio $d/\Lambda$, that is the air-filling fraction of the photonic crystal cladding, increases, successive higher-order modes become trapped [1.5]. A proper geometry design of the fiber cross-section thus guarantees that only the fundamental mode is guided. More detailed studies of the properties of triangular PCFs have shown that this occurs for $d/\Lambda < 0.4$ [1.6, 1.7].

By exploiting this property, it it possible to design very large-mode area fibers, which can be successfully employed for high-power delivery, amplifiers, and lasers. Moreover, by doping the core in order to slightly reduce its refractive index, light guiding can be turned off completely at wavelengths shorter than a certain threshold value.

### 1.2.2 Photonic bandgap guidance

Optical fiber designs completely different form the traditional ones result from the fact that the photonic crystal cladding have gaps in the ranges of the supported modal index $\beta/k_0$ where there are no propagating modes. These are the PBGs of the crystal, which are similar to the two-dimensional bandgaps which characterize planar lightwave circuits, but in this case they have propagation with a non-zero value of $\beta$. It is important to underline that gaps can appear for values of modal index both greater and smaller than unity, enabling the formation of hollow-core fibers with bandgap material as a cladding, as reported in Fig. 1.4. These fibers, which cannot be made using conventional optics, are related to Bragg fibers, since they do not rely on TIR to guide light. In fact, in order to guide light by TIR, it is necessary a lower-index cladding material surrounding the core, but there are no suitable low-loss materials with a refractive index lower than air at optical frequencies [1.1]. The first PCF which exploited the PBG effect to guide light was reported in 1998 [1.5, 1.8], and it

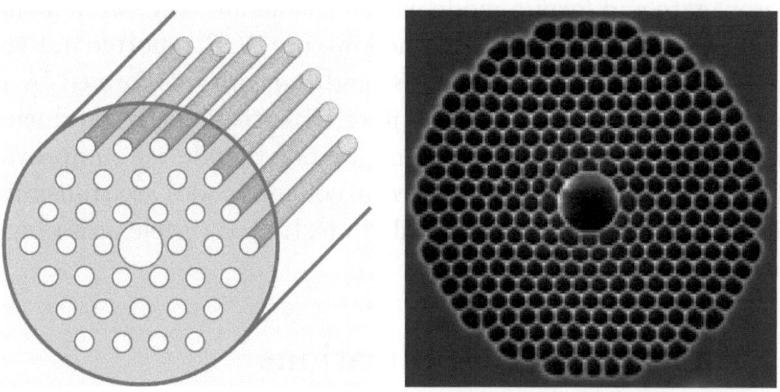

(a) (b)

Figure **1.4**: (a) Schematic of a hollow-core PCF with a triangular lattice of air-holes, which guides light through the photonic bandgap effect. (b) Microscope picture of a fabricated hollow-core triangular PCF, kindly provided by Crystal Fiber A/S.

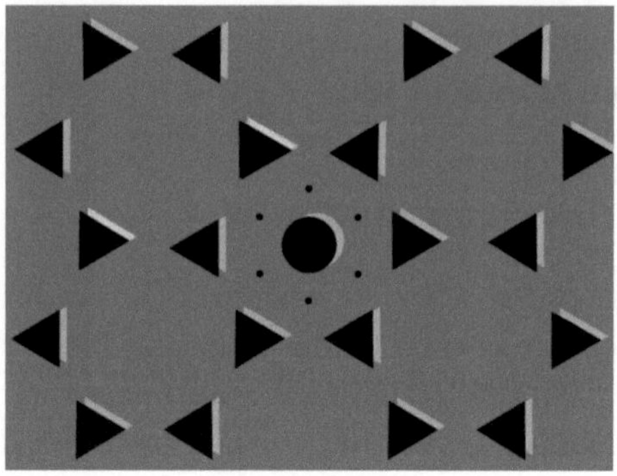

Figure **1.5**: Schematic of the cross-section of the first photonic bandgap PCF with a honeycomb air-hole lattice, proposed in [1.8].

is shown in Fig. 1.5. Notice that its core is formed by an additional air-hole in a honeycomb lattice. This PCF could only guide light in silica, that is in the higher-index material.

Hollow-core guidance had to wait until 1999, when the PCF fabrication technology had advanced to the point where larger air-filling fractions, required to achieve a PBG for air-guiding, became possible [1.5]. Notice that an air-guided mode must have $\beta/k_0 < 1$, since this condition guarantees that light is free to propagate and form a mode within the hollow core, while being unable to escape into the cladding. The first hollow-core PCF, reported in Fig. 1.2, had a simple triangular lattice of air-holes, and the core was formed by removing seven capillaries in the center of the fiber cross-section. By producing a relatively large core, the chances of finding a guided mode were improved. When white light is launched into the fiber core, colored modes are transmitted, thus indicating that light guiding exists only in restricted wavelength ranges, which coincide with the photonic bandgaps [1.5].

## 1.3 Properties and applications

Due to the huge variety of air-holes arrangements, PCFs offer a wide possibility to control the refractive index contrast between the core and the photonic crystal cladding and, as a consequence, novel and unique optical properties.

Since PCFs provide new or improved features, beyond what conventional optical fibers offer, they are finding an increasing number of applications in ever-widening areas of science and technology.

### 1.3.1 Solid-core fibers

Index-guiding PCFs, with a solid glass region within a lattice of air-holes, offer a lot of new opportunities, not only for applications related to fundamental fiber optics. These opportunities are related to some special properties of the photonic crystal cladding, which are due to the large refractive index contrast and the two-dimensional nature of the microstructure, thus affecting the birefringence, the dispersion, the smallest attainable core size, the number of guided modes and the numerical aperture and the birefringence.

#### Highly birefringent fibers

Birefringent fibers, where the two orthogonally polarized modes carried in a single-mode fiber propagate at different rates, are used to maintain polarization states in optical devices and subsystems. The guided modes become birefringent if the core microstructure is deliberately made twofold symmetric, for example, by introducing capillaries with different wall thicknesses above and below the core. By slightly changing the air-hole geometry, it is possible to produce levels of birefringence that exceed the performance of conventional birefringent fiber by an order of magnitude. It is important to underline that, unlike traditional polarization maintaining fibers, such as bow tie, elliptical-core or Panda, which contain at least two different glasses, each one with a different thermal expansion coefficient, the birefringence obtainable with PCFs is highly insensitive to temperature, which is an important feature in many applications. An example of the cross-section of a highly birefringent PCF is reported in Fig. 1.6.

#### Dispersion tailoring

The tendency for different light wavelengths to travel at different speeds is a crucial factor in the telecommunication system design. A sequence of short light pulses carries the digitized information. Each of these is formed from a spread of wavelengths and, as a result of chromatic dispersion, it broadens as it travels, thus obscuring the signal. The magnitude of the dispersion changes with the wavelength, passing through zero at 1.3 μm in conventional optical fibers.

In PCFs, the dispersion can be controlled and tailored with unprecedented freedom. In fact, due to the high refractive index difference between silica

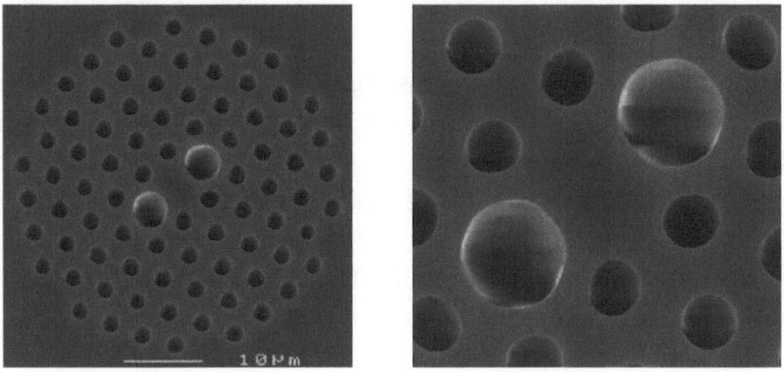


(a) (b)

Figure **1.6**: Microsope picture of (a) the cross-section and (b) the core region of a highly birefringent triangular PCF, kindly provided by Crystal Fiber A/S.

(a) (b)

Figure **1.7**: Microscope picture of (a) the cross-section and (b) the core region of a highly nonlinear PCF, characterized by a small-silica core and large air-holes, with zero-dispersion wavelength shifted to the visible. The pictures are kindly provided by Crystal Fiber A/S.

and air, and to the flexibility of changing air-hole sizes and patterns, a much broader range of dispersion behaviors can be obtained with PCFs than with standard fibers.

For example, as the air-holes get larger, the PCF core becomes more and more isolated, until it resembles an isolated strand of silica glass suspended by six thin webs of glass, as it is shown in Fig. 1.7. If the whole structure is made very small, the zero-dispersion wavelength can be shifted to the visible, since the group velocity dispersion is radically affected by pure waveguide dispersion.

On the contrary, very flat dispersion curves can be obtained in certain wavelength ranges in PCFs with small air-holes, that is with low air-filling fraction. As an example, a dispersion-flattened triangular PCF with seven air-hole rings, characterized by $\Lambda \simeq 2.5\,\mu\text{m}$ and $d \simeq 0.5\,\mu\text{m}$, has been presented in [1.9].

### Ultrahigh nonlinearities

An attractive property of solid-core PCFs is that effective index contrasts much higher than in conventional optical fibers can be obtained by making large air-holes, or by reducing the core dimension, so that the light is forced into the silica core. In this way a strong confinement of the guided-mode can be reached, thus leading to enhanced nonlinear effects, due to the high field intensity in the core. Moreover, a lot of nonlinear experiments require specific dispersion properties of the fibers. As a consequence, PCFs can be successfully exploited to realize nonlinear fiber devices, with a proper dispersion, and this is presently one of their most important applications.

An important example is the so-called supercontinuum generation, that is the formation of broad continuous spectra by the propagation of high power pulses through nonlinear media, as it will be widely described in Section 4.1. The term supercontinuum does not indicate a specific phenomenon, but rather a plethora of nonlinear effects, which, in combination, lead to extreme spectral broadening. The determining factors for supercontinuum generation are the dispersion of the nonlinear medium relative to the pumping wavelength, the pulse length and the peak power. Since the nonlinear effects involved in the spectral broadening are highly dependent on the medium dispersion, a proper design of the dispersion properties can significantly reduce the power requirements. The widest spectra, in fact, can be obtained when the pump pulses are launched close to the zero-dispersion wavelength of the nonlinear media.

### Large-mode area fibers

By changing the geometric characteristics of the fiber cross-section, it is possible to design PCFs with completely different properties, that is with large effective area. The typical cross-section of this kind of fibers, called large mode area (LMA) PCFs, consists of a triangular lattice of air-holes where the core is defined by a missing air-hole. An example of a triangular PCF is reported in Fig. 1.3. The PCF core diameter can be defined as $d_{core} = 2\Lambda - d$, which corresponds to the distance between opposite air-hole edges in the core region. When $d/\Lambda < 0.4$, the triangular PCF is endlessly single mode, that is, single mode at any wavelength [1.6, 1.7]. In this condition it is the core size or the pitch that determines the zero-dispersion wavelength $\lambda_0$, the mode field diameter (MFD) and the numerical aperture (NA) of the fiber. LMA PCFs are usually exploited for high-power applications, since fiber damage and nonlinear limitations are drastically reduced. In particular, LMA fibers are currently used for applications at short wavelengths, that is in ultraviolet (UV) and visible bands, like the generation and delivery of high-power optical beams for laser welding and machining, optical lasers, and amplifiers, providing significant advantages with respect to traditional optical fibers [1.10].

Conventional active fibers for lasers and amplifiers are basically standard transmission fibers whose core region has been doped with rare earth elements. These fibers, also known as "core-pumped," are usually pumped with single-mode pump lasers. Due to its power limitations, this kind of fiber is unsuitable for high-power applications, on the order of 1 W, and upwards. High-power fibers are usually designed with a double-cladding structure, where a second low-index region acts as a cladding for a large pump core. In the center of the pump core is located a much smaller doped signal core, as reported in Fig. 1.8a.

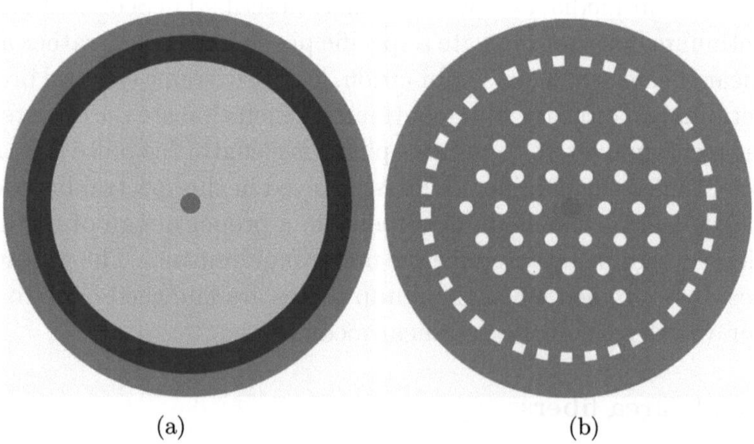

(a) (b)

Figure **1.8**: Schematic of the cross-sections of (a) a standard step-index double-clad fiber and of (b) an air-clad PCF, where the single-mode active core is embedded in a silica-air LMA structure.

With respect to the more traditional core-pumped design, double-cladding fibers present a large pump area and high numerical aperture, thus enabling pumping with relatively low-cost multimode diodes and diode bars/stacks. However, it is important to underline that, when considering high powers, it is necessary to optimize fiber characteristics, such as NA, core dimension, and length, in order to obtain efficient coupling of the pump light, reduction of nonlinear effects, high conversion of pump light and good thermal properties [1.10].

Since PCF structures can provide single-mode waveguides with MFD values above 40 μm, LMA PCFs can be successfully used as active fibers for high-power applications. The PCF double-clad equivalent is shown in Fig. 1.8b. It consists of a LMA structure with a doped-core inside an air-clad pump guide. Very high NA values, determined by the silica bridge width, shown in the three fiber cross-sections reported in Fig. 1.9, are provided by the air-clad, since the refractive index contrast is greatly enhanced. As a consequence, the NA is only limited by the practical handling of the fibers, being the cleaving increasingly challenging at NA values above 0.6. Moreover, the thermal conductivity is greatly improved compared to conventional polymer-clad fibers, because the PCF is made only of glass and air, and there is no material degradation. The power density limit is set only by the silica damage threshold. Finally, the combination of very large MFD and high NA offered by PCFs makes it possible to fabricate lasers and amplifiers with very short fiber lengths, drastically reducing the nonlinear effects [1.10]. As an example, an air-clad $Yb^{3+}$-doped fiber characterized by a hexagonal inner cladding with

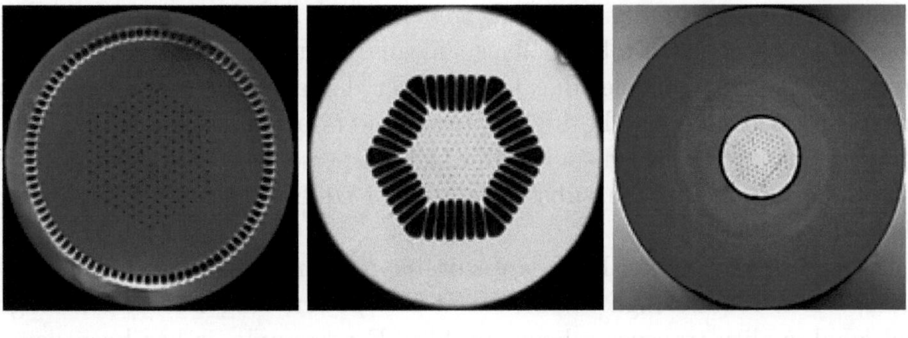

Figure **1.9**: Microscope pictures of the cross-section of three different air-clad PCFs, kindly provided by Crystal Fibre A/S.

a diameter of 117 μm and a NA of about 0.6 has been fabricated by Crystal Fibre A/S [1.10].

For industrial material processing applications, kW average power levels are desirable. These power levels can be now obtained with fiber lasers. By exploiting the advantages offered by air-clad active PCFs, that is large $Yb^{3+}$-doped core mode-field areas and high NA all-silica pump core, reliable $kW$ lasers can be realized with short fiber lengths [1.10].

Once reached the power limit of the fibers previously described, multi-core PCF designs can be exploited in order to obtain a further scaling of the power level. An example is given by the an air-clad $Yb$-doped fiber with seven cores, each with a mode-field diameter of 15 μm fabricated by Crystal Fibre A/S [1.10]. This fiber has been applied in a laser configuration and provided lasing in a supermode with high beam quality. The next planned generation of multicore fibers will have 18 cores [1.10].

### 1.3.2 Hollow-core fibers

Hollow-core PCFs have great potential, since they exhibit low nonlinearity [1.11] and high damage threshold [1.12–1.14], thanks to the air-guiding in the hollow core and the resulting small overlap between silica and the propagating mode. As a consequence, they are good candidates for future telecommunication transmission systems.

Another application, perhaps closer to fruition, which can successfully exploit these advantages offered by air-guiding PCFs, is the delivery of high-power continuous wave (CW), nanosecond and sub-picosecond laser beams, which are useful for marking, machining and welding, laser-Doppler velocimetry, laser surgery, and THz generation [1.15]. In fact, optical fibers would be the most suitable delivery means for many applications, but at present they are unusable, due to the fiber damage and the negative non-linear effects caused by the high optical powers and energies, as well as to the fiber group-velocity dispersion, which disperses the short pulses [1.15]. These limitations can be substantially relieved by considering hollow-core fibers [1.15].

Moreover, air-guiding PCFs are suitable for nonlinear optical processes in gases, which require high intensities at low power, long interaction lengths and good-quality transverse beam profiles. For example, it has been demonstrated that the threshold for stimulated Raman scattering in hollow-core fibers filled with hydrogen is orders of magnitude below that obtained in previous experiments [1.16]. In a similar way, PCFs with a hollow core can be used for trace gas detection or monitoring, or as gain cells for gas lasers.

Finally, the delivery of solid particles down a fiber by using optical radiation pressure has been demonstrated [1.5]. In particular, only 80 mW of a 514 nm argon laser light was enough to levitate and guide 5 µm polystyrene spheres along a 15 cm length of PCF with a hollow-core diameter of 20 µm [1.17].

## 1.4 Loss mechanisms

The most important factor for any optical fiber technology is loss. Losses in conventional optical fibers have been reduced over the last 30 years, and a further improvement is unlikely to be reached. The minimum loss in fused silica, which is around 1550 nm, is slightly less than 0.2 dB/km. This limit is important, since it sets the amplifier spacing in long-haul communications systems, and thus is a major cost of a long-haul transmission system [1.1].

Loss mechanisms in PCFs are here described in details, in order to understand how far the technology can go to reduce their values.

### 1.4.1 Intrinsic loss

#### Solid-core fibers

The optical loss $\alpha_{\mathrm{dB}}$, measured in dB/km, of PCFs with a sufficiently reduced confinement loss, which will be described in Section 1.4.2, can be expressed as

$$\alpha_{\mathrm{dB}} = A/\lambda^4 + B + \alpha_{\mathrm{OH}} + \alpha_{\mathrm{IR}} \ , \tag{1.1}$$

being $A$, $B$, $\alpha_{\mathrm{OH}}$, and $\alpha_{\mathrm{IR}}$ the Rayleigh scattering coefficient, the imperfection loss, and OH and infrared absorption losses, respectively. At the present time the losses in PCFs are dominated by $OH$-absorption loss and imperfection loss [1.18].

In a typical PCF the OH-absorption loss is more than 10 dB/km at 1380 nm and this causes an additional optical loss of 0.1 dB/km in the wavelength range around 1550 nm. Since this contribution is very similar to the intrinsic optical loss of 0.14 dB/km for pure silica glass at this wavelength, the OH-absorption loss reduction becomes an important and challenging problem. Most of the OH impurities seem to penetrate the PCF core region during the fabrication process. As a consequence, a dehydration process is useful in reducing the OH-absorption loss [1.18].

Imperfection loss, caused mainly by air-hole surface roughness, is another serious problem. In fact, during the fabrication process, the air-hole surfaces can be affected by small scratches and contamination. If this surface roughness is comparable with the considered wavelength, it can significantly increase the scattering loss. Thus, it is necessary to improve the polishing and etching process, in order to reduce the optical loss caused by this roughness. Moreover, fluctuation in the fiber diameter during the fiber drawing process can cause an additional imperfection loss, if the air-hole size and pitch change along the fiber [1.18].

It is important to underline that the Rayleigh scattering coefficient of PCFs is the same as that of a conventional SMF. However, this is higher than that of a pure silica-core fiber, although the PCF is made of pure silica glass. It is necessary to reduce the roughness further, in order to obtain a lower imperfection loss and a lower Rayleigh scattering coefficient [1.18].

It is fundamental to fabricate long PCFs with low loss, if they are to be used as transmission media. In Fig. 1.10, the decrease of the loss for fabri-

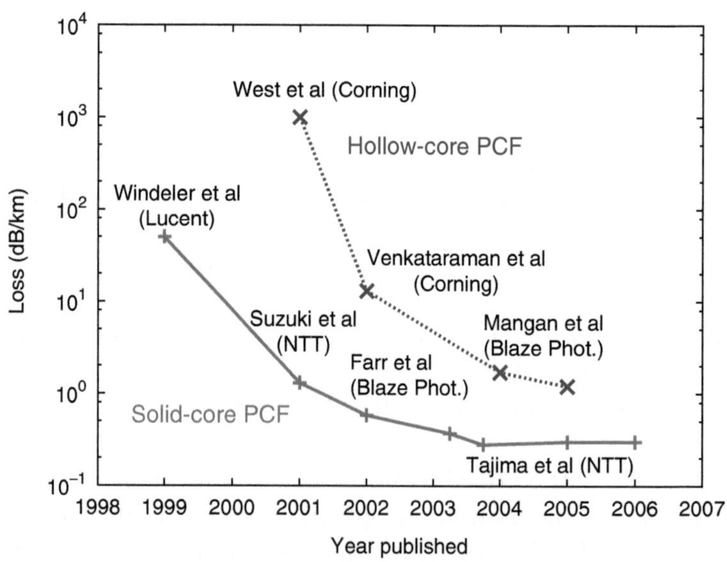


Figure **1.10**: Optical loss behavior during the last years, until 2006, for solid-core (+ symbols) and hollow-core ($\times$ symbols) PCFs.

cated index-guiding PCFs is shown until 2006. Early in their development, solid-core PCFs had optical losses of the order of 0.24 dB/m [1.19] and the available length was limited to several meters. The optical losses of PCFs were rapidly reduced to 1 dB/km by improving the fabrication process [1.20–1.22]. The lowest loss yet achieved is 0.28 dB/km [1.23]. Recently, a low loss, that is 0.3 dB/km, and long length, that is 100 km, PCF was reported [1.24]. The optical losses of these kind of PCFs are still high compared with those of a conventional SMF. However, a solid-core PCF is not expected to have significantly lower losses than standard fibers.

### Hollow-core fibers

Losses in hollow-core fibers are limited by the same mechanisms as in conventional fibers and in index-guiding PCFs, that is absorption, Rayleigh scattering, confinement loss, bend loss, and variations in the fiber structure along the length. However, there is the possibility to reduce them below the levels found in conventional optical fibers, since the majority of the light travels in the hollow core, in which scattering and absorption could be very low.

As shown in Fig. 1.10, the attenuation values reported in literature for hollow-core PCFs are higher than those for both solid-core PCFs and standard fibers. Looking at the attenuation profiles for a range of hollow-core fibers made by Crystal Fibre A/S, reported in [1.10], it is possible to notice two important facts. The guiding bandwidth is usually around 15% of the central wavelength and the loss scales inversely with the wavelength. As indicated by theoretical considerations, the attenuation related to mode coupling and scattering at the internal air–silica interfaces should scale with the wavelength $\lambda$ as $\lambda^{-3}$ [1.25]. This consideration has been confirmed by experimental observations [1.26] and applies to the seven-cell design, that is to fibers whose hollow core has been obtained by removing seven capillaries in the center of the fiber cross-section. It is important to underline that, in order to reach lower losses, 19-cell designs can be used, as it is demonstrated by the loss values reported in Fig. 1.11 [1.26]. The minimum loss of 1.7 dB/km has been obtained with the hollow-core PCF shown in Fig. 1.12 [1.26], since the larger core reduces the overlap of the guided modes with silica. Recently, a record attenuation as low as 1.2 dB/km at 1620 nm has been reported with the same kind of fiber [1.25]. However, notice that a larger hollow core gives increased perimeters, leading to a greater density of surface modes, which will be described in the following, leading to decreased bandwidth and also to increased higher-order dispersion [1.27].

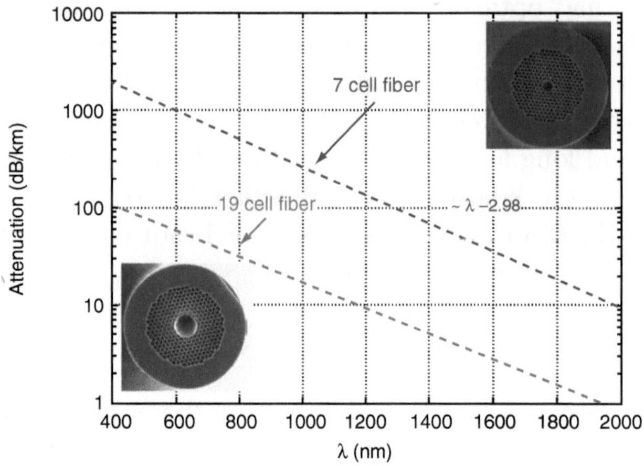


Figure **1.11**: Attenuation behavior versus the wavelength for seven-cell and 19-cell hollow-core PCFs, whose cross-sections are shown as inset. The microscope pictures of the hollow-core fibers are kindly provided by Crystal Fiber A/S.

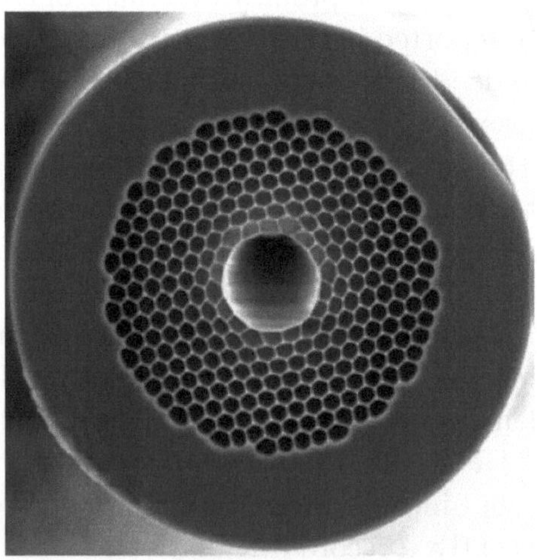

Figure **1.12**: Microscope picture of a 19-cell hollow-core fiber, kindly provided by Crystal Fiber A/S.

Reducing the hollow-core PCF loss to levels below those of conventional silica fibers remains a challenge. As it will be discussed later, confinement losses can be eliminated by forming a photonic crystal cladding with a sufficient number of air-hole rings, while bending losses, which are determined by the fiber design, can be reduced to a low level, at least in some structures. For what concerns Rayleigh scattering, as well as absorption, they should be reduced to below the level in bulk fibers, even if the increased scattering at the many surfaces represents potentially a problem. However, the biggest unknown is the level of variation in the fiber structure along its length. In fact, the bandgap presents a high sensitivity to structural fluctuations that occur over long fiber lengths, that is wavelengths that are guided in one section may leak away in another.

It is possible to reduce the fiber nonuniformity with a more careful fabrication process, but not to eliminate the surface roughness due to *surface capillary waves* (SCWs) frozen into the fiber when it is made. In fact, SCWs, which exist on liquid surfaces, such as molten glass, where surface tension provides a restoring force, freeze as the glass solidifies, leaving a surface roughness given by the SCW amplitudes when the temperature equals the glass transition one [1.25]. This roughness scatters light from the fundamental mode to the not guided ones, thus causing the fiber loss. It has been demonstrated [1.25, 1.28] that this surface roughness ultimately limits the hollow-core PCF attenuation. In fact, due to its thermodynamic origin, this roughness is not reduced with a better fiber drawing process. Other technological improvements in homogeneity are likely to reduce the attenuation of hollow-core PCFs by no more than a factor of two [1.25, 1.28]. Moreover, the negative effect of the roughness can be decreased by a proper hollow-core fiber design, that is by reducing the overlap of the fundamental mode with the glass–air surfaces. By acting in these two directions, the hollow-core PCF attenuation can plausibly be reduced from the actual record of 1.2 dB/km at 1620 nm to 0.2 dB/km at the same wavelength [1.25, 1.28]. Some further improvements can derive from the choice of a longer operating wavelength, since the scattering loss decreases. On the contrary, the infrared absorption loss increases because some light propagates in the glass. By considering the $\lambda^{-3}$ behavior of the attenuation previously described, and by estimating the guided-mode absorption in the glass lower than 1%, it has been shown [1.25, 1.28] that the plausible loss of a hollow-core PCF could fall to about 0.13 dB/km if optimized for 1900 nm, as reported in Fig. 1.13.

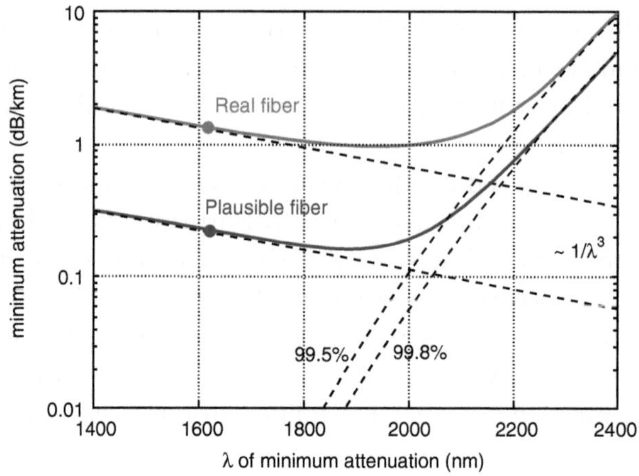


Figure **1.13**: Minimum attenuation extrapolation of the real hollow-core PCF with 1.2 dB/km at 1620 nm, and of a plausible perfected one with loss of 0.2 dB/km at 1620 nm, as described in [1.25, 1.28].

In order to further reduce the attenuation, new fiber designs, new materials or a method for increasing surface tension are required [1.25, 1.28].

Finally, there is an excess loss in hollow-core PCFs which occurs at wavelengths where there is coupling from the air-guided fundamental mode to the confined *surface modes*, which have much greater overlap with the glass and, as a consequence, experience far higher loss.

The presence of surface modes strongly affects the guiding properties of the air-guiding PCFs by reducing the width of their transmission window [1.29]. For example, it has been demonstrated [1.30] that the attenuation spectrum of the hollow-core fiber characterized by a hole-to-hole spacing of 4.7 μm, an air-filling fraction of 0.94 and a core diameter of about 12.7 μm presents a high-loss region in the wavelength range between 1550 and 1650 nm, which is due to the surface mode presence. This loss, which affects the modes confined in the hollow core, is caused by the surface modes through the coupling to the core modes, as well as to the lossy extended ones.

In particular, air-guiding fibers, like conventional ones, are characterized by a finite number of guided modes, all comprised in the hollow core, and by an infinite number of leaky cladding and radiation modes. Ideally, being the core modes completely confined in air, the small perurbations in the silica structure cause only a slight coupling with the cladding ones, which have the largest spatial overlap in the perturbed region [1.30]. However, in some hollow-core PCFs another kind of core modes, called surface modes, has been found, which are not unexpected in periodic structures [1.30]. In fact, surface modes occur when an infinite photonic crystal is abruptly terminated, since these modes satisfy the new set of boundary conditions introduced by the terminations, where they are localized [1.31]. Moreover, the surface mode presence in the periodic structure strongly depends on the termination location. For example, they are induced in photonic crystals made by dielectric rods in air only if the termination cross the rods [1.31]. In a similar way, in air-guiding PCFs the core defect introduces a pertubation in the lattice periodicity. Differently from the well-known case of a planar interface of a semi-infinite periodic structure, in hollow-core fibers there is a finite circular interface between the free space and the periodic structure, that is the surface defined by the core radius. The surface modes, which decay exponentially in both the periodic structure and the free space, are supported in this region. It is important to underline that the surface modes are localized near the core, being their wavelength within the PBG, but they differ from the truely guided modes since most of their intensity is confined in the silica which surrounds the hollow core. As shown in Fig. 1.14, the surface mode presence significantly affects the behavior of

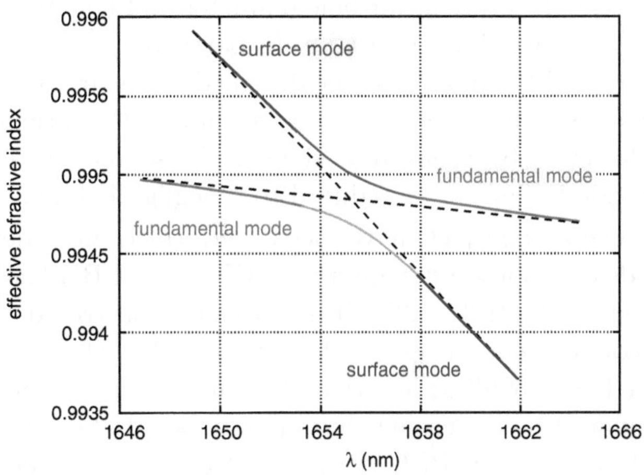


Figure **1.14**: Example of a hollow-core PCF dispersion curve with an avoided crossing between the fundamental mode and a surface mode.

the fundamental mode dispersion curve, with anticrossing points between the guided mode, confined in the core, and the surface one. A reduction of the silica quantity around the hollow core causes an energy increase of the surface modes, which consequently move into the PBG, while, on the contrary, the guided-mode energy is unaffected by this change. However, for some hollow-core dimensions there is a certain interaction between these two different kinds of modes, due to the significant field overlap in the silica regions. Thus, the axial perturbations all along the fiber can provide the light coupling between the core and the surface modes. The loss mechanism related to surface modes is complete by considering that they are highly overlapped and coupled with the continuum of the extended modes in the cladding [1.30].

### 1.4.2 Confinement loss

In both solid-core and hollow-core PCFs it is necessary to consider another contribution to the losses, that is the leakage or confinement losses. These are due to the finite number of air-holes which can be made in the fiber cross-section. As a consequence, all the PCF guided modes are leaky.

For example, in solid-core PCFs light is confined within a core region by the air-holes. Light will move away from the core if the confinement provided by the air-holes is inadequate. This means that it is important to design such aspects of the PCF structure as air-hole diameter and hole-to-hole spacing, or pitch, in order to realize low-loss PCFs. In particular, the ratio between the air-hole diameter and the pitch must be designed to be large enough to confine light into the core. On the other hand, a large value of the ratio makes the PCF multi-mode. However, by properly designing the structure, the confinement loss of single-mode PCFs can be reduced to a negligible level.

Recently, several analyses have been performed in order to find the guidelines to design both index-guiding PCFs and PBG-based fibers with negligible leakage losses [1.32–1.37]. It has been demonstrated a strong dependence of the confinement losses on the number of air-hole rings, especially for fibers with high air-filling fraction. In particular, leakage losses can be significantly reduced by increasing the ring number [1.36]. Finally, simulation results have shown that in PBG fibers the leakage loss dependence on the number of air-hole rings is much weaker than in index-guiding PCFs, whereas the confinement losses exhibit a strong dependence on the position of the localized state inside the PBG [1.33].

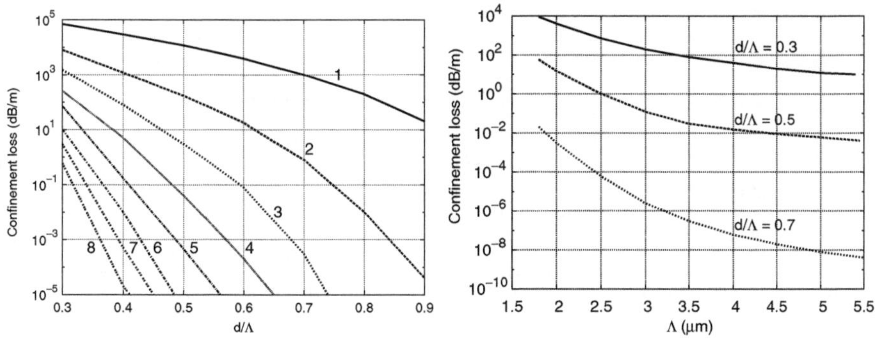


(a) (b)

Figure **1.15**: Leakage loss at 1550 nm (a) as a function of the air-hole diameter $d$ normalized to the pitch $\Lambda = 2.3$ μm for different ring numbers and (b) as a function of the pitch $\Lambda$ for different air-filling fraction $d/\Lambda$ [1.32].

In order to better explain the leakage loss behavior in PCFs, a solid-core fiber and a hollow-core one with a triangular lattice of air-holes are here considered. The silica-core one, represented in Fig. 1.3, has $\Lambda = 2.3$ μm [1.32]. As shown in Fig. 1.15a, its leakage loss, calculated according to Eq. (A.9) as explained in Appendix A, quickly decreases when the air-hole ring number or the air-hole diameter increases. The reduction rate of the confinement loss increases in the same way with these geometric parameters. As expected, the loss decreases with larger $\Lambda$ values for a fixed $d/\Lambda$. In this case, $\Lambda$ and $d$ are scaled in the same way, so a larger pitch corresponds to a larger silica core size and, as a consequence, to a higher field confinement. The wavelength dependence of the confinement loss is shown in Fig. 1.16 for two different pitch values, that is 2.3 and 4.6 μm. Since the field becomes less confined, the leakage loss increases with $\lambda$. Moreover, the ring number affects the wavelength dependence, which is weaker for few air-hole rings [1.32]. The triangular hollow-core fiber taken as the second PCF example, shown in Fig. 1.4, is characterized by $d = 1.8$ μm and $\Lambda = 2$ μm. Fig. 1.17 reports the wavelength dependence of the confinement loss when four and seven air-hole rings are considered in the fiber cladding. In both cases, the leakage loss sprectrum presents a U-shape with a minimum value around the normalized wavelength $\lambda/\Lambda = 0.68$, which corresponds to the central position of the guided mode inside the PBG. When the defect state moves close to the PBG edges, the loss increases more quickly

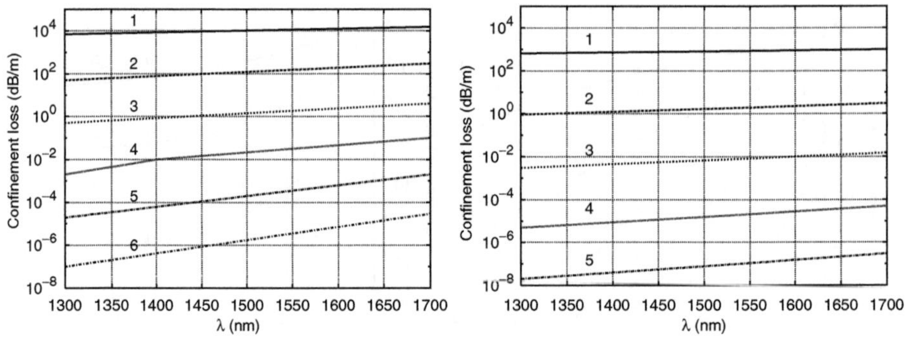


(a) (b)

Figure **1.16**: Leakage loss as a function of the wavelength $\lambda$ for different ring numbers, $d/\Lambda = 0.5$ and (a) $\Lambda = 2.3$ μm and (b) $\Lambda = 4.6$ μm [1.32].

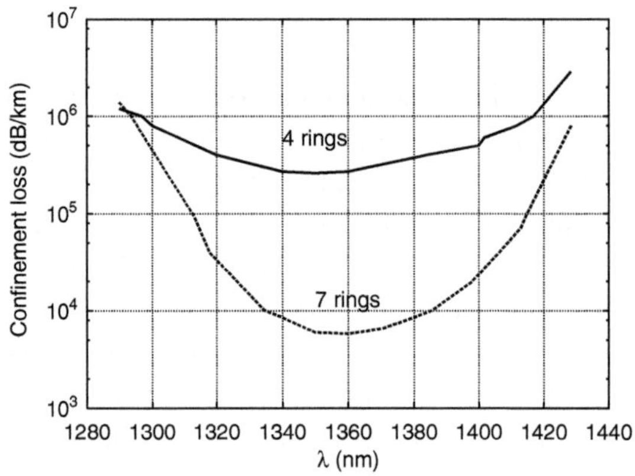


Figure **1.17**: Confinement loss versus the wavelength in a triangular hollow-core PCF with four and seven air-hole rings [1.33].

when the air-hole ring number is higher. Despite the high air-filling fraction, that is $d/\Lambda = 0.9$, the dependence on the ring number is very weak, if compared with the case of solid-core PCFs. Finally, it is important to underline that there is a strong wavelength dependence of the loss. For example, the loss of the seven air-hole ring PCF increases of a decade with respect to the minimum value in a wavelength range of less than 100 nm [1.33].

### 1.4.3 Bending loss

As already stated, an alternative route to fabricate LMA fibers is offered by PCFs, which can be designed to be endlessly single-mode, unlike conventional fibers that exhibit a cutoff wavelength below which higher-order modes are supported. As for standard optical fibers, the practical achievable mode area in PCFs is limited by the macrobending loss [1.38–1.40].

Conventional fibers suffer additional loss if bent more tightly than a certain critical radius. For wavelengths longer than a certain value, that is the "long-wavelength bend loss edge," all guidance is effectively lost. The same behavior is observed also in PCFs, which show even a "short-wavelength bend loss edge" [1.41], caused by bend-induced coupling from the fundamental to the higher-order modes, which leak out of the core. In fact, at short wavelengths the guided mode is mainly confined into the silica [1.41] and when $\lambda << \Lambda$ the field can escape through the interstitial space between the neighboring air-holes. As a consequence, the fiber becomes more sensitive to bending.

By considering triangular PCFs, shown in Fig. 1.3, a large air-filling fraction, that is a high $d/\Lambda$ value, results in a better resistance to the bending loss, whereas the hole-to-hole spacing $\Lambda$ roughly determines the position of the minimum of the bending loss curve, which roughly occurs at $\Lambda/2$ [1.39]. Since LMA PCFs are generally designed with $\Lambda > 7 - 8$ μm, the standard telecommunication window falls in the short-wavelength edge. In spite of that, it has been demonstrated that LMA PCFs exhibit bending losses comparable with those of similarly sized conventional fibers at 1550 nm [1.38, 1.42–1.45]. Moreover, it has been experimentally shown that PCFs optimized for visible applications are more robust towards bending at any of the wavelengths from 400 to 1000 nm compared to a conventional fiber which is single-mode at the visible wavelengths [1.46].

PCFs with larger relative air-hole diameters, that is with higher $d/\Lambda$, are less sensitive to bending loss. However, the demand for single-mode operation and the need for large-mode size limits the increase of $d/\Lambda$, and other solutions must be adopted. It has been demonstrated that the bending losses of triangular PCFs can be improved by changing the air-hole configuration from the traditional single-rod core design [1.47, 1.48]. In particular, an alternative structure with the core region formed by three silica rods has been proposed, with the aim to improve the guided-mode area and the resistance to the bending loss, particularly at the short wavelengths [1.47]. An accurate evaluation of the advantages regarding the bending loss that can be obtained by comparing suitably matched three-rod and single-rod PCFs designs has been recently performed [1.48]. Numerical results have shown that, when the silica core is formed with three adjacent rods, the critical bending radius, defined as the radius at which the loss equals 3 dB/loop, can be reduced by approximately 20% with respect to the traditional single-rod PCF designs at 1064 nm, in excellent agreement with the experimental measurements.

Many different approaches have been proposed in literature to evaluate the bending loss in conventional optical fibers, which usually assume a circular symmetric refractive index profile. Unfortunately, these approaches are not straightforward in PCFs, due to the complex nature of their refractive index profile. As a consequence, an accurate modeling of bending loss becomes even more challenging. A theoretical model that successfully predicts the bending loss in LMA PCFs is described in [1.44], where the physical origin of the phenomenon is investigated, accounting for two different mechanisms that contribute to the overall loss, that is transition loss and pure bend loss [1.43, 1.44]. The transition loss occurs where the curvature of the fiber changes suddenly, that is at the beginning or the end of the bend. This loss can be modeled as a sort of coupling loss, because the mode fields in the straight and curved sections are not aligned. The pure bend loss is the continuous loss that occurs along any curved section of fiber, due to the inability of the tails of the field to keep in phase with the faster-travelling central portion of the field. In this model, the full refractive index profile of the PCF is retained and hence the six fold field shape as well. In fact, the bent fiber is modeled as a straight fiber with an equivalent index profile, given by a transformation that superimposes a gradient onto the refractive index of the straight fiber in the direction of the bend. Other theoretical approaches have been developed, which provide a correct parametric dependence of the bending loss with the PCF geometric parameters [1.39, 1.40, 1.49]. An analogy with the conventional step-index optical fibers has been applied, by introducing an effective normalized frequency for the PCFs, with an equivalent core radius and an effective refractive index for the microstructured cladding [1.39]. Then, in order to describe the PCF bending loss, an expression for the power loss coefficient of the standard optical fibers due to the macrobending is considered [1.39]. As reported in [1.50], this semianalytical effective-index model correctly predict the short-wavelength loss behavior measured in a triangular PCF with $d = 2.4$ μm and $\Lambda = 7.8$ μm, that is with $d/\Lambda \simeq 0.31$. On the contrary, the difference between the bending loss values measured and numerically evaluated is significant for triangular PCFs with a higher air-filling fraction.

For example, there is no agreement between simulation and measurement results for the triangular fiber with $d = 5.5$ μm and $\Lambda = 10$ μm, even if a full vectorial effective index calculation is performed, since triangular PCFs with these geometric characteristics are not strictly single-mode [1.50]. Recently, an easy-to-evaluate expression for the bending loss prediction in triangular LMA PCFs has been proposed [1.49]. The validity of the expression, which is based on a recent formulation of the V-parameter for PCFs [1.7], has been experimentally verified for different fiber geometric parameters and bending diameters. As reported in [1.49], it has been demonstrated that the difference between the bend-loss edge measured and numerically predicted is within the uncertainty of the measurements.

Hollow-core PCFs have different bending properties with respect to silica-core ones. For applications like high-power delivery for medical use or material processing, which are suitable for air-guiding fibers, a low bending sensitivity is required, since it allows a very flexible use and an easy integration in supporting mechanical systems [1.51]. After an early demonstration in a theoretical work of the low influence of bending on the hollow-core PCF guiding properties [1.52], the bending loss of air-guiding fibers have been experimentally measured [1.51, 1.53]. In particular, a single-mode fiber and a multi-mode one have been considered for the experimental measurements, which have indicated that no significant effect can be observed even by applying 10 turns with a small bending diameter of 4 cm [1.51,1.53]. The most important effect obtained with bending is a shift of the short-wavelength bandgap edge towards longer wavelengths, thus causing a PBG narrowing for the hollow-core PCFs. On the contrary, a similar shift has not been measured at the long-wavelength bandgap edge. In order to understand the fact that air-guiding PCFs are bending insensitive over most of the PBG, it is useful to consider the difference between the refractive index of the core, that is 1, and of the PBG edge, which corresponds to the cladding one. Being this difference very high, that is about $2 \cdot 10^{-2}$, a very tight confinement of the guided-mode in the hollow-core can be obtained, which results in the robust guiding even through tightly bent PCFs [1.53].

## 1.5 Fabrication process

One of the most important aspect in designing and developing new fibers is their fabrication process. In a lot of papers presented in literature so far, PCFs have been realized by “introducing” air-holes in a solid glass material. This has several advantages, since air is mechanically and thermally compatible with most materials, it is transparent over a broad spectral range, and it has a very low refractive index at optical frequencies. Fibers fabricated using silica and air have been accurately analyzed, partly because most conventional optical fibers are produced from fused silica. This is also an excellent material to work with, because viscosity does not change much with temperature and it is relatively cheap. Moreover, filling the holes of a silica–air structure opens up a wide range of interesting possibilities, such as the bandgap guidance in a low-index core made of silica when the holes are filled with a high-index liquid.

Traditional optical fibers are usually manufactured by fabricating a fiber preform and drawing it with a high-temperature furnace in a tower setup [1.54]. The different vapor deposition techniques, for example, the modified chemical vapor deposition (MCVD), the vapor axial deposition (VAD), and the outside vapour deposition (OVD), are all tailored for the fabrication of circular-symmetric fiber preforms. Thus, the deposition can be controlled in a very accurate way only in the radial direction without significant modifications of the methods. Moreover, producing conventional single-mode optical fibers requires core and cladding materials with similar refractive index values, which typically differ by around 1%, and are usually obtained by vapor deposition techniques. On the contrary, designing PCFs requires a far higher refractive index contrast, differing by perhaps 50–100% [1.1]. As a consequence, all the techniques previously described are not directly applicable to the fabrication of preforms for microstructured optical fibers, whose structure is not characterized by a circular symmetry.

Differently from the drawing process of conventional optical fibers, where viscosity is the only really important material parameter, several forces are important in the case of PCFs, such as viscosity, gravity, and surface tension. This is due to the much larger surface area in a microstructured geometry, and to the fact that many of the surfaces are close to the fiber core, thus making surface tension relatively much more important. As a consequence, the choice of the base material strongly influences the technological issues and applications in the PCF fabrication process.

### 1.5.1 Stack-and-draw technique

In order to fabricate a PCF, it is necessary, first, to create a preform, which contains the structure of interest, but on a macroscopic scale. One possibility to exploit for the PCF fabrication is the drilling of several tens to hundreds

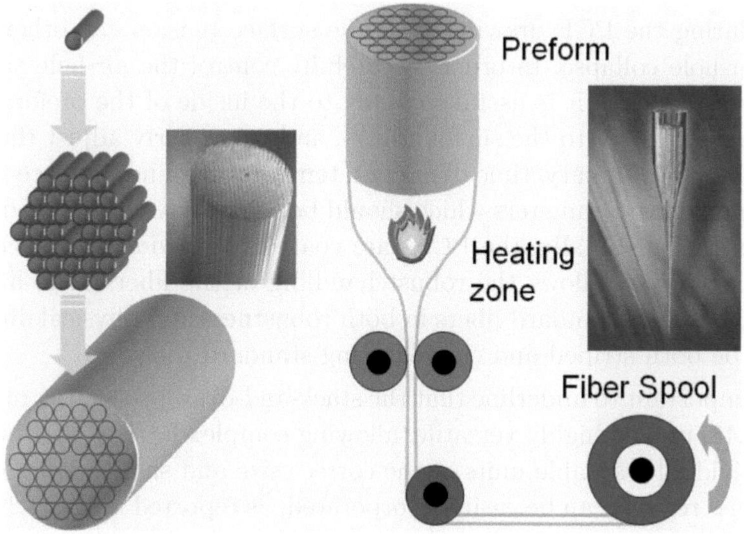


Figure **1.18**: Scheme of the PCF fabrication process. The photographs of the PCF preforms have been kindly provided by Crystal Fiber A/S.

of holes in a periodic arrangement into one final preform. However, a different and relatively simple method, called stack-and-draw, introduced by Birks et al. in 1996 [1.55], has become the preferred fabrication technique in the last years, since it allows relatively fast, clean, low-cost, and flexible preform manufacture.

The PCF preform is realized by stacking by hand a number of capillary silica tubes and rods to form the desired air–silica structure, as reported in Fig. 1.18. This way of realizing the preform allows a high level of design flexibility, since both the core size and shape, as well as the index profile throughout the cladding region can be controlled. After the stacking process, the capillaries and rods are held together by thin wires and fused together during an intermediate drawing process, where the preform is drawn into preform canes. This intermediate step is important in order to provide numerous preform canes for the development and optimization of the later drawing of the PCFs to their final dimensions [1.54]. Then, the preform is drawn down on a conventional fiber-drawing tower, greatly extending its length, while reducing its cross-section, from a diameter of 20 mm to a 80–200 μm one, as shown in Fig. 1.18. With respect to standard optical fibers, which are usually drawn at temperatures around 2100°C, a lower temperature level, that is 1900°C, is kept during the PCF drawing since the surface tension can otherwise lead to the air-hole collapse. In order to carefully control the air-hole size during the drawing process, it is useful to apply to the inside of the preform a slight overpressure relative to the surroundings, and to properly adjust the drawing speed [1.54]. In summary, time dynamics, temperature, and pressure variations are all significant parameters which should be accurately controlled during the PCF fabrication. Finally, the PCFs are coated to provide a protective standard jacket, which allows the robust handling of the fibers. The final PCFs are comparable to standard fibers in both robustness and physical dimensions, and can be both striped and cleaved using standard tools.

It is important to underline that the stack-and-draw procedure, represented in Fig. 1.18, proved highly versatile, allowing complex lattices to be assembled from individual stackable units of the correct size and shape. Solid, empty, or doped glass regions can be easily incorporated, as reported in Fig. 1.19. A wide range of different structures have been made by exploiting this technique, each with different optical properties. Moreover, overall collapse ratios as large as about 50,000 times have been realized, and continuous holes as small as 25 nm in diameter have been demonstrated, earning an entry in the Guinness Book of Records in 1999 for the World's Longest Holes [1.5].

A very important issue is the comparison of the PCF stack-and-draw procedure with the vapor deposition methods usually employed for standard optical fibers. Obviously, it is more difficult that the preforms for conventional optical fibers become contaminated, since their surface area is smaller. Moreover, the

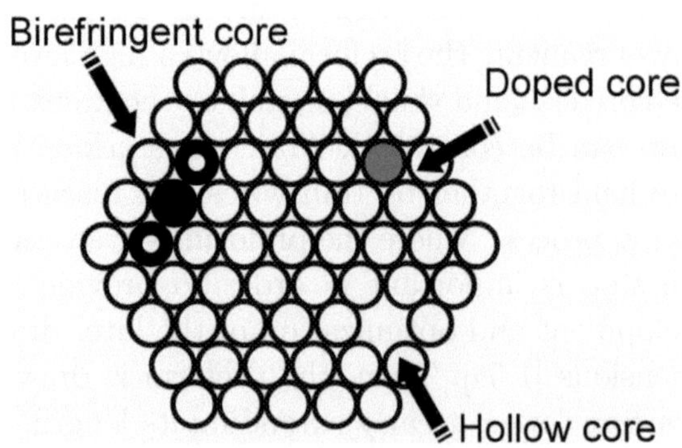


Figure **1.19**: Example of a PCF cross-section, showing the flexibility offered by the stack-and-draw fabrication process.

stacking method requires a very careful handling, and the control of air-hole dimensions, positions, and shapes in PCFs makes the drawing significantly more complex [1.54]. Finally, it is important to underline that the fabrication process of PCFs with a hollow core, realized by removing some elements from the stack center, is much more difficult than that of standard optical fibers, even if at present fibers with low loss and practical lengths have been obtained [1.56].

### 1.5.2 Extrusion fabrication process

Silica–air preforms have also been extruded, enabling the formation of structures not readily attainable by stacking capillaries [1.1]. The extrusion process has been recently applied to other glasses, which are not as readily available in tube form as silica, like compound glasses. These materials, which provide a lot of interesting properties, like an extended wavelength range for transmission and higher values of the nonlinear coefficient, can be used to fabricate preforms through the extrusion process due to their lower softening temperatures, which make easier the fabrication procedure [1.54].

In this fabrication process a molten glass is forced through a die containing a suitably designed pattern of holes. Extrusion allows fiber to be drawn directly from bulk glass, using a fiber-drawing tower, and almost any structure, crystalline or amorphous, can be produced. It works for many materials, including chalcogenides, polymers, and compound glasses. However, selective doping of specified regions, in order to introduce rare earth ions or render the glass photosensitive, is much more difficult.

Different PCFs produced by the extrusion process have been presented in literature. In particular, the fabrication of the first non-silica glass PCF by exploiting this technique has been reported in 2002 by Kiang et al. [1.57]. A commercial glass, called Schott SF57 glass, has been used, which has a softening temperature of only 519°C and a high lead concentration, which causes a relatively high refractive index of 1.83 at a wavelength of 633 nm and of 1.80 at 1530 nm. This material is interesting since its nonlinear refractive index, that is $4.1 \cdot 10^{-19}$ $W^2/m$ at 1060 nm, is more than one order of magnitude larger than that of pure silica. Another highly nonlinear PCF has been fabricated with the bismuth-oxide-based glass, which has proved to be an attractive novel material for nonlinear devices and compact $Er^{3+}$-doped amplifiers [1.58]. The fiber fabrication presented in [1.58] consists of three steps. In the first step, the structured preform of 16 mm outer diameter and the jacket tube are extruded.

In the second step, the preform is reduced in scale on a fiber-drawing tower to a cane of about 1.6 mm diameter. In the last step, the cane is inserted within the jacket tube, and this assembly is drawn down to the final fiber.

Extrusion has been also used to fabricate a highly nonlinear PCFF with SF6, a commercial glass produced by Schott, which has a refractive index of 1.76 at 1550 nm and a nonlinear index $n_2 = 2.2 \times 10^{19}$ $m^2/W$, higher than that of silica [1.59]. Starting from the preform, fibers of tens of meter lengths with core diameters in the range 1–10 μm have been drawn [1.59].

Notice that the method proposed in [1.59] can also be applied to other commercial glasses, including some with higher nonlinearity and slightly lower intrinsic loss. In particular, a tellurite PCF with an outer diameter of 190 μm and a core diameter of 7 μm has been realized, as described in [1.60].

Recently, a PCF with the highest value of nonlinearity yet reported for an optical fiber, that is 1860 $(W \cdot km)^{-1}$ at 1550 nm, and improved loss values has been fabricated by extrusion with a three-step procedure using the Schott SF57 glass [1.61]. The three-step procedure used for the highly nonlinear PCF fabrication is shown in Fig. 1.20a, while a schematic of the cross-section of the extruded PCF is reported in Fig. 1.20b. By applying the same fabrication approach to other glass materials with nonlinearity higher than that of SF57 glass, it will be possible to fabricate fibers with even higher values of the effective nonlinearity per unit length [1.56].

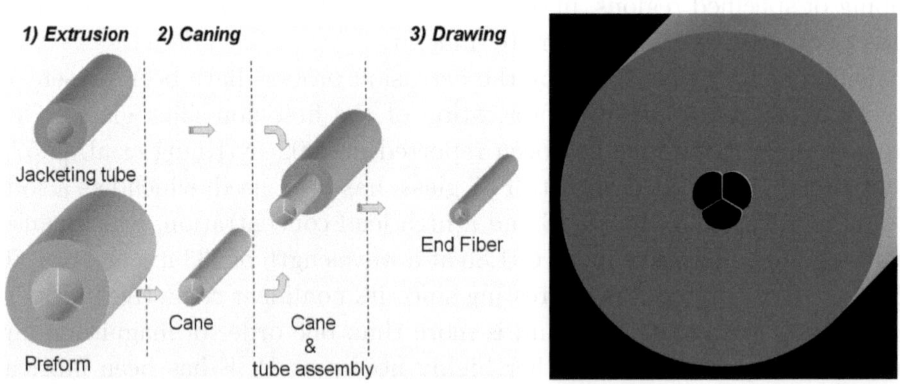


(a) (b)

Figure **1.20**: (a) Scheme of the fabrication process of the extruded SF57 glass PCF and (b) schematic of the cross-section of the fiber proposed in [1.56].

### 1.5.3 Microstructured polymer optical fibers

Microstructured polymer optical fiber (MPOF) have been also fabricated and presented for the first time in 2001 [1.62]. The light-guiding mechanism in MPOFs is the same as in PCFs, since it arises from a pattern of microscopic air-holes which run all along the fiber length. MPOFs have emerged as a viable alternative to glass PCFs for specific applications, due to the relatively low draw temperatures associated with polymers, usually polymethyl methacrylate (PMMA). A range of different materials and fabrication methods can be used to make MPOF preforms. In addition to the capillary stacking technique, traditionally used for glass PCFs, polymer preforms can be made using techniques such as extrusion, polymerization in a mold, drilling or injection molding. With such techniques available, it becomes straightforward to obtain different cross-sections in the preform, with air-holes of arbitrary shapes and sizes in any desired arrangement [1.63].

The material properties of PMMA provide advantages relative to silica in the fabrication of PCFs, because the drawing of all these fibers is governed by the balance between surface tension and viscosity-related forces. While the viscosity of PMMA and silica are of similar magnitudes at their respective draw temperatures, PMMA surface tension is an order of magnitude lower than that of silica. Thus, by lowering the draw temperature, and hence increasing both the viscosity and the required draw tension, air-hole distortion and collapse due to surface tension effects can be minimized, allowing fine-scale MPOFs to be drawn.

The overall MPOF fabrication procedure is presented in [1.64]. After designing the structure required in the final fiber and taking into account the expected 30–40% hole collapse during fabrication, the air-hole pattern is drilled into the primary preform using a computer numerical-controlled mill. As reported in [1.65], the coated drill bits produce deep holes with minimal drill wander, while leaving the inside of the holes with a smooth finish, the latter being of importance in that it minimizes the likelihood of surface roughness induced scattering in the drawn fiber. Air-hole sizes at the preform stage are typically 1–10 mm in diameter. At the present time, the finest primary preform structure that can be drilled involves 1 mm holes with 0.1 mm wall thickness between holes to a depth of 65 mm. The longest preform that can currently be drilled is 140 mm in length, using 2 mm drills that are 70 mm long with a hole spacing of 2.5 mm. Note that the air-holes are drilled from both ends of the preform [1.64].

Primary preforms can be drawn directly to fiber using a one-stage process, that is with primary oven only, although the fiber diameter control is generally poor. The main role of the primary oven is thus to produce either a "stretched" secondary preform or a microstructured cane which is subsequently sleeved to form a secondary preform. The alternative employed depends primarily on the dimensions required for the air-hole structure in the final fiber. For most MPOF designs, the stretched secondary preform is drawn directly to fiber. However, some MPOF designs, such as the small-core fibers, require that the final air-hole sizes be of the order of a micron, or less, and hence the sleeving technique is used. The final step involves drawing the secondary preform to fiber [1.64].

In one of the realized fiber presented in [1.63] the photonic crystal cladding consists of four rings of air-holes in a triangular lattice, embedded in an outer sleeve. Small deformations are present, such as in the air-hole diameters and shapes. Moreover, compared to the preform that the fiber was drawn from, the air-hole structure in the fiber has a slightly reduced $d/\Lambda$ ratio [1.63].

Casting is another useful technique for fabricating the MPOF preforms, whose low-cost mass production is an important issue [1.66]. This method offers some advantages with respect to stack-and-draw and drill-machining, since it is possible to change easily the mold structure in order to make preforms of different kinds of MPOF, with particular shape, dimension, and disposition of the air-holes [1.66]. Moreover, the chemical and physical contamination, which can cause the optical inhomogeneity, can be significantly reduced if casting is realized in a sealed vessel, thus improving the MPOF scattering loss performances [1.66].

In order to make preforms by casting, which can be used for both glass and polymer, chemical precursors, like monomer, initiator, and chain-transfer agent, are introduced in a mold which mirrors the desired air-hole distribution in the preform. Then, after the polymer setting in the mold, the solid structure is completely removed [1.66]. Even if a mold is expensive to design and produce, it becomes an economic solution for a large production, because it is used to fabricate many thousands of molded items. The molds used in the process described in [1.66] are usually made of alloy stainless steel with a smooth and highly polished surface, and are formed by separate parts, so that they can be opened and the molded items can be easily removed. The final preform structure is defined in the mold through the presence of steel wires or rods, which are usually releasably attached in order to allow an individual removal.

The first MPOF preform fabrication by casting has been demonstrated in 2001 [1.67], when a fiber with four rings of air-holes organized in a triangular lattice has been drawn from a preform of 50 mm diameter and 250 mm length. More recently, casting has been applied to make the preform of a LMA MPOF [1.68] and to fabricate a PMMA preform with a large diameter, that is 7 cm, characterized by 88 air-holes in a triangular lattice and an overall length of 40 cm [1.66]. This fiber preform has been made with a mold formed by a glass tube, 88 metal rods, which define the air-holes, and two Teflon plates, used to keep fixed the rods. The casted preform 40 cm long has been used to fabricate more than a hundred kilometer of MPOF. The final fiber, drawn in a three-stage process, which is necessary due to the large size of the preform, is highly birefringent. In fact, it is characterized by an elliptical core obtained by removing three air-holes [1.66].

### 1.5.4 OmniGuide fibers

A unique cigar-rolling technique has also been reported for a polymer–glass combination [1.69]. In this technique, a multilayer mirror is effectively rolled up to form a preform with a hollow core. This structure differs from the others reported above in several respects. In fact, it uses two solid materials, but in a configuration that results in an almost exclusively radial variation in refractive index, as shown in the schematic reported in Fig. 1.21. The radial-only index variation has intrinsic advantages for forming hollow-core fibers, being the PBG-based guiding in a hollow core much easier to create, in principle, because only a single periodicity is involved. On the contrary, the use of two solid materials limits the choice to those with compatible thermal and thermo-mechanical properties.

Excellent progress has been demonstrated to date, and a structure with a bandgap in the 10 μm wavelength band has been produced [1.69]. In this fiber the hollow core is surrounded by a solid multilayer structure of high-refractive index contrast, leading to large photonic bandgaps and omnidirectional reflectivity. In order to achieve high index contrast in the layered portion of the fiber, a chalcogenide glass with a refractive index of about 2.8, that is arsenic triselenide $As_2Se_3$, has been combined with a high glass-transition temperature thermoplastic polymer, having a refractive index of about 1.55, that is polyether sulfone (PES). The same polymer was used as a cladding material, resulting in fibers composed of about 98% polymer by volume, not

Figure **1.21**: Schematic of the cross-section of a hollow-core cylindrical multilayer fiber.

including the hollow core. These fibers, called OmniGuide fibers, thus combine high optical performance with polymeric processability and mechanical flexibility. A variety of hollow-core fibers have been realized by depositing an $As_2Se_3$ layer, which is 5–10 μm thick, by thermal evaporation onto a 25–50 μm-thick PES film, and the subsequent "rolling" of that coated film into a hollow multilayer tube, called a fiber preform. This hollow macroscopic preform was consolidated by heating under vacuum, and clad with a thick outer layer of PES. The layered preform was then placed in an optical fiber draw tower, and drawn down into tens or hundreds of meters of fiber having well-controlled submicrometer layer thicknesses [1.69].

## 1.6 Photonic crystal fibers in the market

PCFs have always attracted a strong interest among the researchers since 1996. In fact, the microstructure presence in the optical fiber cross-section has provided enhanced physical performances, which have led to new developments in different application areas [1.70].

During the last decade the development of PCFs has been strongly driven by the academia searching for new exciting waveguiding principles, and, at the same time, by the interest of large companies, such as Lucent Technologies,

Corning and NTT, which have focused parts of their resources on this new class of specialty fibers [1.54]. The establishment of start-up companies, like Crystal Fibres A/S and Blaze Photonics Ltd. (now a part of Crystal Fibres A/S, Bath, UK), has been the result of the academic activities developed over the last 7 years [1.54]. These companies have fabricated a lot of new PCF products for the reasearch market, getting a strong patent portfolio. Due to the presence of such different players, it is now difficult to predict which kind of enterprise will eventually dominate the PCF market [1.54].

In the last decade, the research field of PCFs, which was initially a revolutionary discovery, has become a mature technology, with many types of products with a variety of unique properties fabricated and sold by different companies all over the world. For example, different hollow-core fibers with the transmission bandwidth centered at 800, 1060 and 1550 nm are commercialy available in the market. Moreover, the PCF attenuation has become more and more close to that of conventional optical fibers, and a high level of fabrication reproducibility has been reached [1.70]. Even if most of the actual PCF customers are still university research groups, this situation is likely to change in the next future.

At the moment, it can be expected that PCFs for higher-power next-generation fiber lasers and amplifiers, and for supercontinuum generation will be the first products to reach the "real" market, that is to gain commercial opportunities also outside the academic world [1.54, 1.70]. In fact, even if PCFs were originally envisioned as a solution for higher data rates in telecommunications, conventional optical fibers currently in use are so good that PCFs do not offer an obvious advantage right now.

For the future, the most interesting possibilities for PCFs are related to fiber-based signal-processing devices with tunable properties, fibers for dispersion management, and gas- or liquid-filled fiber-based sensor devices [1.71]. In the meanwhile, an intense research in the PCF field should continue in both academic research centers and companies, in order to obtain a further reduction of the losses, in particular for hollow-core fibers, which can have an important role for future telecommunications, to investigate alternative materials, and to expand the range of possible PCF designs and applications.

# Bibliography

[1.1] J. C. Knight, "Photonic crystal fibres," *Nature*, vol. 424, pp. 847–851, Aug. 2003.

[1.2] P. St. J. Russell and R. Dettmer, "A neat idea [photonic crystal fibre]," *IEE Review*, vol. 47, pp. 19–23, Sept. 2001.

[1.3] J. C. Knight, T. A. Birks, P. St. J. Russell, and D. M. Atkin, "Pure silica single-mode fibre with hexagonal photonic crystal cladding," in *Proc. Optical Fiber Communications Conference OFC 1996*, San Jose, California, USA, Feb. 25 – Mar. 1, 1996.

[1.4] R. F. Cregan, B. J. Mangan, J. C. Knight, T. A. Birks, P. St. J. Russell, P. J. Roberts, and D. C. Allan, "Single-mode photonic band gap guidance of light in air," *Science*, vol. 285, pp. 1537–1539, Sept. 1999.

[1.5] P. St. J. Russell, "Photonic crystal fibers," *Science*, vol. 299, pp. 358–362, Jan. 2003.

[1.6] B. T. Kuhlmey, R. C. McPhedran, C. M. de Sterke, P. A. Robinson, G. Renversez, and D. Maystre, "Microstructured optical fibers: where's the edge?" *Optics Express*, vol. 10, pp. 1285–1290, Nov. 2002. Available at: http://www.opticsexpress.org/abstract.cfm?URI=OPEX-10-22-1285

[1.7] N. A. Mortensen, J. R. Folkenberg, M. D. Nielsen, and K. P. Hansen, "Modal cutoff and the V parameter in photonic crystal fibers," *Optics Letters*, vol. 28, pp. 1879–1881, Oct. 2003.

[1.8] J. C. Knight, J. Broeng, T. A. Birks, and P. St. J. Russell, "Photonic band gap guidance in optical fibers," *Science*, vol. 282, pp. 1476–1478, Nov. 1998.

[1.9] W. H. Reeves, J. C. Knight, P. St. J. Russell, and P. J. Roberts, "Demonstration of ultra-flattened dispersion in photonic crystal fibers," *Optics Express*, vol. 10, pp. 609–613, July 2002. Available at: http://www.opticsexpress.org/abstract.cfm?URI=OPEX-10-14-609

[1.10] R. E. Kristiansen, K. P. Hansen, J. Broeng, P. M. W. Skovgaard, M. D. Nielsen, A. Petersson, T. P. Hansen, B. Mangan, C. Jakobsen, and H. R. Simonsen, "Microstructured fibers and their applications," in *Proc. Reuniòn Española de Optoelectrònica OPTOEL 2005*, Elche, Spain, July 13–15, 2005.

[1.11] D. G. Ouzounov, F. R. Ahmad, D. Müller, N. Venkataraman, M. T. Gallagher, M. G. Thomas, J. Silcox, K. W. Koch, and A. L. Gaeta, "Generation of megawatt optical solitons in hollow-core photonic bandgap fibers," *Science*, vol. 301, pp. 1702–1704, Sept. 2003.

[1.12] C. J. S. de Matos, J. R. Taylor, T. P. Hansen, K. P. Hansen, and J. Broeng, "All-fiber chirped pulse amplification using highly-dispersive air-core photonic bandgap fiber," *Optics Express*, vol. 11, pp. 2832–2837, Nov. 2003. Available at: http://www.opticsexpress.org/abstract.cfm?URI=OPEX-11-22-2832

[1.13] J. Limpert, T. Schreiber, S. Nolte, H. Zellmer, and A. Tünnermann, "All fiber chirped-pulse amplification system based on compression in air-guiding photonic bandgap fiber," *Optics Express*, vol. 11, pp. 3332–3337, Dec. 2003. Available at: http://www.opticsexpress.org/abstract.cfm?URI=OPEX-11-24-3332

[1.14] C. J. S. de Matos and J. R. Taylor, "Chirped pulse Raman amplification with compression in air-core photonic bandgap fiber," *Optics Express*, vol. 13, pp. 2828–2834, Apr. 2005. Available at: http://www.opticsexpress.org/abstract.cfm?URI=OPEX-13-8-2828

[1.15] G. Humbert, J. C. Knight, G. Bouwmans, P. St. J. Russell, D. P. Williams, P. J. Roberts, and B. J. Mangan, "Hollow core photonic crystal fibers for beam delivery," *Optics Express*, vol. 12, pp. 1477–1484, Apr. 2004. Available at: http://www.opticsexpress.org/abstract.cfm?URI=oe-12-8-1477

[1.16] F. Benabid, J. C. Knight, G. Antonopoulos, and P. St. J. Russell, "Stimulated Raman scattering in hydrogen-filled hollow-core photonic crystal fiber," *Science*, vol. 298, pp. 399–402, Oct. 2002.

[1.17] F. Benabid, J. C. Knight, and P. St. J. Russell, "Particle levitation and guidance in hollow-core photonic crystal fiber," *Optics Express*, vol. 10, pp. 1195–1203, Oct. 2002. Available at: http://www.opticsexpress.org/abstract.cfm?URI=OPEX-10-21-1195

[1.18] K. Kurokawa, K. Tajima, K. Tsujikawa, and K. Nakajima, "Reducing the losses in photonic crystal fibres," in *Proc. European Conference on Optical Communication ECOC 2005*, Glasgow, Scotland, Sept. 25–29, 2005.

[1.19] P. J. Bennett, T. M. Monro, and D. J. Richardson, "Toward practical holey fiber technology: fabrication, splicing, modeling, and characterization," *Optics Letters*, vol. 24, pp. 1203–1205, Sept. 1999.

[1.20] K. Tajima, K. Nakajima, K. Kurokawa, N. Yoshizawa, and M. Ohashi, "Low-loss photonic crystal fibers," in *Proc. Optical Fiber Communications Conference OFC 2002*, Anaheim, California, USA, Mar. 17–22, 2002, pp. 523–524.

[1.21] L. Farr, J. C. Knight, B. J. Mangan, and P. J. Roberts, "Low loss photonic crystal fiber," in *Proc. European Conference on Optical Communication ECOC 2002*, Copenhagen, Denmark, Sept. 8–12, 2002, paper PD1.3.

[1.22] K. Tajima, J. Zhou, K. Nakajima, and K. Sato, "Ultra low loss and long length photonic crystal fiber," in *Proc. Optical Fiber Communications Conference OFC 2003*, Atlanta, Georgia, USA, Mar. 23–28, 2003, pp. PD1–1–PD1–3.

[1.23] K. Tajima, J. Zhou, K. Kurokawa, and K. Nakajima, "Low water peak photonic crystal fibers," in *Proc. European Conference on Optical Communication ECOC 2003*, Rimini, Italy, Sept. 21–25, 2003, paper Th4.1.6.

[1.24] K. Kurokawa, K. Tajima, J. Zhou, K. Nakajima, T. Matsui, and L. Sankawa, "Penalty-free dispersion-managed soliton transmission over 100 km low loss PCF," in *Proc. Optical Fiber Communications Conference OFC 2005*, Anaheim, California, USA, Mar. 6–11, 2005.

[1.25] P. J. Roberts, F. Couny, H. Sabert, B. J. Mangan, D. Williams, L. Farr, M. Mason, A. Tomlinson, T. A. Birks, J. C. Knight, and P. St. J. Russell, "Ultimate low loss of hollow-core photonic crystal fibres," *Optics Express*, vol. 13, pp. 236–244, Jan. 2005. Available at: http://www.opticsexpress.org/abstract.cfm?URI=oe-13-1-236

[1.26] B. J. Mangan, L. Farr, A. Langford, P. J. Roberts, D. P. Williams, F. Couny, M. Lawman, M. Mason, S. Coupland, R. Flea, H. Sabert, T. A. Birks, J. C. Knight, and P. St. J. Russell, "Low loss (1.7 dB/km) hollow core photonic bandgap fiber," in *Proc. Optical Fiber Communications Conference OFC 2004*, Anaheim, California, USA, Feb. 23–27, 2004, paper PDP24.

[1.27] J. C. Knight, "Optical fibres using microstructured optical materials," in *Proc. European Conference on Optical Communication ECOC 2005*, Glasgow, Scotland, Sept. 25–29, 2005, paper We3.1.

[1.28] T. A. Birks, "Reducing losses in photonic crystal fibres," in *Proc. Optical Fiber Communications Conference OFC 2006*, Anaheim, California, USA, Mar. 5–10, 2006.

[1.29] R. Amezcua-Correa, N. G. Broderick, M. N. Petrovich, F. Poletti, and D. J. Richardson, "Optimizing the usable bandwidth and loss through core design in realistic hollow-core photonic bandgap fibers," *Optics Express*, vol. 14, pp. 7974–7985, Aug. 2006. Available at: http://www.opticsexpress.org/abstract.cfm?URI=oe-14-17-7974

[1.30] J. A. West, C. M. Smith, N. F. Borrelli, D. C. Allan, and K. W. Koch, "Surface modes in air-core photonic band-gap fibers," *Optics Express*, vol. 12, pp. 1485–1496, Apr. 2004. Available at: http://www.opticsexpress.org/abstract.cfm?URI=oe-12-8-1485

[1.31] M. J. F. Digonnet, H. K. Kim, J. Shin, S. Fan, and G. S. Kino, "Simple geometric criterion to predict the existence of surface modes in air-core photonic-bandgap fibers," *Optics Express*, vol. 12, pp. 1864–1872, May 2004. Available at: http://www.opticsexpress.org/abstract.cfm?URI=oe-12-9-1864

[1.32] D. Ferrarini, L. Vincetti, M. Zoboli, A. Cucinotta, and S. Selleri, "Leakage properties of photonic crystal fibers," *Optics Express*, vol. 10, pp. 1314–1319, Nov. 2002. Available at: http://www.opticsexpress.org/abstract.cfm?URI=OPEX-10-23-1314

[1.33] L. Vincetti, D. Ferrarini, M. Zoboli, A. Cucinotta, F. Poli, and S. Selleri, "Leakage losses in photonic band gap fibers," in *Proc. European Conference on Optical Communication ECOC 2003*, Rimini, Italy, Sept. 21–25, 2003.

[1.34] T. P. White, R. C. McPhedran, C. M. de Sterke, L. C. Botten, and M. J. Steel, "Confinement losses in microstructured optical fibers," *Optics Letters*, vol. 26, pp. 1660–1662, Nov. 2001.

[1.35] V. Finazzi, T. M. Monro, and D. J. Richardson, "Confinement loss in highly nonlinear holey optical fibres," in *Proc. Optical Fiber Communications Conference OFC 2002*, Anaheim, California, USA, Mar. 17–22, 2002, paper ThS4.

[1.36] D. Ferrarini, L. Vincetti, M. Zoboli, A. Cucinotta, F. Poli, and S. Selleri, "Leakage Losses in Photonic Crystal Fibers," in *Proc. Optical Fiber Communications Conference OFC 2003*, Atlanta, Georgia, USA, Mar. 23–28, 2003, paper FI5.

[1.37] K. Saitoh and M. Koshiba, "Confinement Losses in air-guiding photonic bandgap fibers," *IEEE Photonics Technology Letters*, vol. 15, pp. 236–238, Feb. 2003.

[1.38] J. C. Baggett, T. M. Monro, K. Furusawa, and D. J. Richardson, "Comparative study of large-mode holey and conventional fibers," *Optics Letters*, vol. 26, pp. 1045–1047, July 2001.

[1.39] T. Sørensen, J. Broeng, A. Bjarklev, E. Knudsen, and S. E. B. Libori, "Macro-bending loss properties of photonic crystal fibre," *Electronics Letters*, vol. 37, pp. 287–289, Mar. 2001.

[1.40] N. A. Mortensen and J. R. Folkenberg, "Low-loss criterion and effective area considerations for photonic crystal fibers," *Journal of Optics A: Pure and Applied Optics*, vol. 5, pp. 163–167, May 2003.

[1.41] J. C. Knight, T. A. Birks, R. F. Cregan, P. St. J. Russell, and J. P. de Sandro, "Large mode area photonic crystal fibre," *Electronics Letters*, vol. 34, pp. 1347–1348, June 1998.

[1.42] T. M. Monro, J. C. Baggett, K. Furusawa, and D. J. Richardson, "Comparative Study of Bend Loss in Large Mode Holey and Conventional Fibres," in *Proc. Conference on Lasers and Electro-Optics CLEO 2001*, Baltimore, USA, May 6–11, 2001, p. 259.

[1.43] J. C. Baggett, T. M. Monro, K. Furusawa, and D. J. Richardson, "Distinguishing transition and pure bend losses in holey fibers," in *Proc.*

*Conference on Lasers and Electro-Optics CLEO 2002*, Long Beach, USA, May 19–24, 2002, p. 49.

[1.44] J. C. Baggett, T. M. Monro, K. Furusawa, V. Finazzi, and D. J. Richardson, "Understanding bending loss in holey optical fibers," *Optics Communications*, vol. 227, pp. 317–335, Nov. 2003.

[1.45] K. Miyake, M. Hachiwaka, T. Kinoshita, S. Yamaguchi, H. Kubota, and S. Kawanishi, "Bend resistant photonic crystal fiber compatible with conventional single mode fiber," in *Proc. European Conference on Optical Communication ECOC 2004*, Stockholm, Sweden, Sept. 5–9, 2004.

[1.46] M. D. Nielsen, J. R. Folkenberg, N. A. Mortensen, and A. Bjarklev, "Bandwidth comparison of photonic crystal fibers and conventional single-mode fibers," *Optics Express*, vol. 12, pp. 430–435, Feb. 2004. Available at: http://www.opticsexpress.org/abstract.cfm?URI=oe-12-3-430

[1.47] N. A. Mortensen, M. D. Nielsen, J. R. Folkenberg, A. Petersson, and H. Simonsen, "Improved large-mode-area endlessly single-mode photonic crystal fibers," *Optics Letters*, vol. 28, pp. 393–395, Mar. 2003.

[1.48] J. C. Baggett, T. M. Monro, J. R. Hayes, V. Finazzi, and D. J. Richardson, "Improving bending losses in holey fibers," in *Proc. Optical Fiber Communications Conference OFC 2005*, Anaheim, California, USA, Mar. 6–11, 2005, paper OWL4.

[1.49] M. D. Nielsen, N. A. Mortensen, M. Albertsen, J. R. Folkenberg, A. Bjarklev, and D. Bonacinni, "Predicting macrobending loss for large-mode area photonic crystal fibers," *Optics Express*, vol. 12, pp. 1775–1779, Apr. 2004. Available at: http://www.opticsexpress.org/abstract.cfm?URI=oe-12-8-1775

[1.50] T. Sørensen, J. Broeng, A. Bjarklev, T. P. Hansen, E. Knudsen, S. E. B. Libori, H. R. Simonsen, and J. R. Jensen, "Spectral macro-bending loss considerations for photonic crystal fibres," *IEEE Proceedings Optoelectronics*, vol. 149, pp. 206–210, Oct./Dec. 2002.

[1.51] T. P. Hansen, J. Broeng, and A. Bjarklev, "Macrobending loss in air-guiding photonic crystal fibres," in *Proc. European Conference on Optical Communication ECOC 2003*, Rimini, Italy, Sept. 21–25, 2003.

[1.52] E. Knudsen, A. Bjarklev, J. Broeng, and S. E. Barkou, "Macro-bending loss estimation for air-guiding photonic crystal fibre," in *Proc. Optical Fibre Sensors Conference OFS-14*, Venice, Italy, Oct. 11–13, 2000.

[1.53] T. P. Hansen, J. Broeng, C. Jakobsen, G. Vienne, H. R. Simonsen, M. D. Nielsen, P. M. W. Skovgaard, J. R. Folkenberg, and A. Bjarklev, "Air-guiding photonic bandgap fibers: spectral properties, macrobending loss, and practical handling," *IEEE/OSA Journal of Lightwave Technology*, vol. 22, pp. 11–15, Jan. 2004.

[1.54] J. Lægsgaard and A. Bjarklev, "Microstructured optical fibers – fundamentals and applications," *Journal of the American Ceramic Society*, vol. 89, pp. 1–12, Jan. 2006.

[1.55] T. A. Birks, D. M. Atkin, G. Wylangowski, P. St. J. Russell, and P. J. Roberts, "2D photonic band gap structures in fibre form," in *Photonic Band Gap Materials*, C. M. Soukoulis (ed.) Dordrecht: Kluwer, 1996, pp. 437–444.

[1.56] D. J. Richardson, F. Poletti, J. Y. Y. Leong, X. Feng, H. Ebendorff-Heidepriem, V. Finazzi, K. E. Frampton, S. Asimakis, R. C. Moore, J. C. Baggett, J. R. Hayes, M. N. Petrovich, M. L. Tse, R. Amezcua, J. H. V. Price, N. G. R. Broderick, P. Petropoulos, and T. M. Monro, "Advances in microstructured fiber technology," in *Proc. IEEE/LEOS Workshop on Fibres and Optical Passive Components WFOPC 2005*, Palermo, Italy, June 22–24, 2005.

[1.57] K. M. Kiang, K. Frampton, T. M. Monro, R. Moore, J. Tucknott, D. W. Hewak, D. J. Richardson, and H. N. Rutt, "Extruded singlemode non-silica glass holey optical fibres," *Electronics Letters*, vol. 38, pp. 546–547, June 2002.

[1.58] P. Petropoulos, H. Ebendorff-Heidepriem, S. Asimakis, R. C. Moore, K. Frampton, F. Koizumi, T. M. Monro, and D. J. Richardson, "Extruded small-core bismuth oxide glass holey fibres," in *Proc.*

*Summer-School on Advanced Glass-Based Nano-Photonics POWAG 2004*, Bath, UK, July 12–16, 2004.

[1.59] V. V. R. K. Kumar, A. K. George, W. H. Reeves, J. C. Knight, P. St. J. Russell, F. G. Omenetto, and A. J. Taylor, "Extruded soft glass photonic crystal fiber for ultrabroad supercontinuum generation," *Optics Express*, vol. 10, pp. 1520–1525, Dec. 2002. Available at: http://www.opticsexpress.org/abstract.cfm?URI=OPEX-10-25-1520

[1.60] V. V. R. K. Kumar, A. K. George, J. C. Knight, and P. St. J. Russell, "Tellurite photonic crystal fiber," *Optics Express*, vol. 11, pp. 2641–2645, Oct. 2003. Available at: http://www.opticsexpress.org/abstract.cfm?URI=OPEX-11-20-2641

[1.61] J. Y. Y. Leong, P. Petropoulos, S. Asimakis, H. Ebendorff-Heidepriem, R. C. Moore, K. Frampton, V. Finazzi, X. Feng, J. H. V. Price, T. M. Monro, and D. J. Richardson, "A lead silicate holey fiber with $\gamma = 1860\ W^{-1}km^{-1}$ at 1550 nm," in *Proc. Optical Fiber Communications Conference OFC 2005*, Anaheim, California, USA, Mar. 6–11, 2005.

[1.62] M. van Eijkelenborg, M. Large, A. Argyros, J. Zagari, S. Manos, N. A. Issa, I. M. Bassett, S. C. Fleming, R. C. McPhedran, C. M. de Sterke, and N. A. P. Nicorovici, "Microstructured polymer optical fibre," *Optics Express*, vol. 9, pp. 319–327, Sept. 2001. Available at: http://www.opticsexpress.org/abstract.cfm?URI=OPEX-9-7-319

[1.63] M. A. van Eijkelenborg, A. Argyros, G. Barton, I. M. Bassett, M. Fellew, G. Henry, N. A. Issa, M. C. Large, S. Manos, W. Padden, L. Poladian, and J. Zagari, "Recent progress in microstructured polymer optical fibre fabrication and characterisation," *Optical Fiber Technology*, vol. 9, pp. 199–209, Oct. 2003.

[1.64] G. Barton, M. A. van Eijkelenborg, G. Henry, M. C. Large, and J. Zagari, "Fabrication of microstructured polymer optical fibres," *Optical Fiber Technology*, vol. 10, pp. 325–335, Oct. 2004.

[1.65] P. St. J. Russell, "Photonic Crystal Fibres," in *Proc. Summer-School on Advanced Glass-Based Nano-Photonics POWAG 2004*, Bath, UK, July 12–16, 2004.

[1.66] Y. Zhang, K. Li, L. Wang, L. Ren, W. Zhao, R. Miao, M. C. J. Large, and M. A. van Eijkelenborg, "Casting preforms for microstructured polymer optical fibre fabrication," *Optics Express*, vol. 14, pp. 5541–5547, June 2006. Available at: http://www.opticsexpress.org/abstract.cfm?URI=oe-14-12-5541

[1.67] J. Choi, D. Y. Kim, and U. C. Paek, "Fabrication and properties of polymer photonic crystal fibers," in *Proc. Plastic Optical Fiber Conference POF 2001*, Amsterdam, Netherlands, Sept. 27–30, 2001.

[1.68] D. Asnaghi, A. Gambirasio, A. Macchetta, D. Sarchi, and F. Tassone, "Fabrication of a large-effective-area microstructured plastic optical fibre: design and transmission test," in *Proc. European Conference on Optical Communication ECOC 2003*, Rimini, Italy, Sept. 21–25, 2003, pp. 632–633.

[1.69] B. Temelkuran, S. D. Hart, G. Benoit, J. D. Joannopoulos, and Y. Fink, "Wavelength-scalable hollow optical fibres with large photonic bandgaps for $CO_2$ laser transmission," *Nature*, vol. 420, pp. 650–653, Dec. 2002.

[1.70] A. Bjarklev, "Photonic crystal fibers: fundamentals to emerging application," in *Proc. Conference on Lasers and Electro-Optics CLEO 2005*, Baltimore, USA, May 22–27, 2005, p. 213.

[1.71] *Corning® Photonic Band Gap Specialty Fibers–Hollow Core Design*, Corning. Available at: http://www.corning.com/photonicmaterials/pdf/PI207_PCF_04-06.pdf

# Chapter 2

# Guiding properties

This Chapter summarizes the results obtained by analyzing the PCF guiding properties. These can be evaluated starting from a parameter which characterizes the PCF modes, that is the value of the complex propagation constant $\gamma = \alpha + jk_0 n_{\text{eff}}$, being $\alpha$ the attenuation constant, $n_{\text{eff}}$ the effective index and $k_0$ the wave number in the vacuum.

First of all, results regarding a new kind of PCFs, with a square lattice of air-holes in a silica matrix, are reported. The influence of the lattice geometric parameters, that is the hole-to-hole spacing, or pitch, $\Lambda$ and the ratio $d/\Lambda$ between the air-hole diameter $d$ and the pitch, on the effective index $n_{\text{eff}}$ of the PCF fundamental mode has been accurately investigated [2.1]. Moreover, the modal cutoff of square-lattice PCFs has been evaluated by taking into account the leakage losses, that is the attenuation constant $\alpha$, according to Eq. (A.9) of the second-order mode [2.2]. Both these analyses, already presented in literature for triangular PCFs, have been performed for PCFs with a square lattice of air-holes.

The same method used for the square-lattice PCFs has been applied to study the cutoff properties of a new kind of LMA triangular PCFs, called seven-rod, which have a large silica core obtained by removing the central air-hole and the ones belonging to the first ring [2.3,2.4]. In fact, it is important to investigate the trade-off between the effective area and the single-mode operation regime of seven-rod triangular PCFs, in order to successfully exploit them in practical applications.

Finally, the guiding, the leakage, and the birefringence properties of modified honeycomb PCFs with a hollow core, which guide light through the PBG effect, are described [2.5–2.7]. Air-guiding in PCFs with this kind of lattice is interesting as the fiber provides a larger PBG across the air-line, defined as $n_{\rm eff} = 1$, with respect to that obtained with the triangular lattice. Moreover, the confinement loss of the fundamental and the first higher-order mode has been calculated for the the modified honeycomb PCFs here designed, in order to evaluate the wavelength range where the hollow-core fibers, highly birefringent or not, can be considered effectively single mode.

## 2.1 Square-lattice PCFs

It has been already underlined in Chapter 1 that PCFs have particular properties, strictly related to the geometric characteristics of the air-holes in their cross-section. As a consequence, it is interesting to analyze how a regular air-hole disposition different from the more common triangular one can affect the characteristics of the guided mode. Moreover, it is important to understand in which terms all the results usually obtained for the triangular PCFs can be applied to fibers with different lattice geometries.

To this aim, a PCF with a square lattice, whose cross-section is shown in Fig. 2.1a, has been considered. In this fiber, the air-holes are organized in a square lattice, characterized by the same geometric parameters as the triangular one, that is $\Lambda$ and $d/\Lambda$. Note that the technological feasibility of

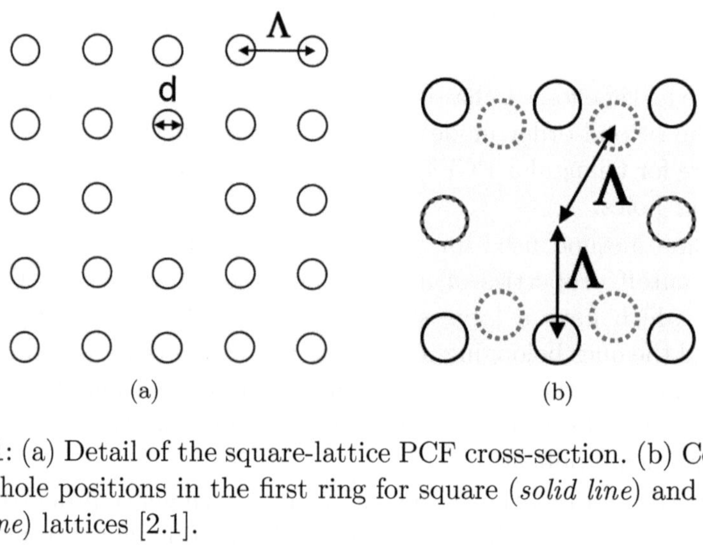


Figure **2.1**: (a) Detail of the square-lattice PCF cross-section. (b) Comparison of the air-hole positions in the first ring for square (*solid line*) and triangular (*dashed line*) lattices [2.1].

square-lattice PCFs has been demonstrated, and they can be drawn from intermediate preforms realized with the standard stack-and-draw fabrication process [2.8]. Recently, square-lattice fibers have been fabricated and characterized in order to analyze their polarization properties, and a great potential for high birefringence has been shown [2.9, 2.10].

### 2.1.1 Guidance

The guiding properties of PCFs with a square lattice of air-holes have been investigated as a function of the geometric characteristics, that is the hole-to-hole spacing $\Lambda$ and the diameter $d$ of the air-hole in the fiber cross-section, as reported in Fig. 2.1a. All the studied square-lattice PCFs have a silica core, obtained by introducing a defect, that is by removing an air-hole, in the center of the fiber transverse section. Fig. 2.1b reports the first ring of air-holes of a square-lattice PCF and a triangular one with the same $\Lambda$ and $d$ values, showing a lower average value of the refractive index around the core in the triangular PCF. In fact, in this case the first ring comprises six air-holes whose distance from the core center is equal to $\Lambda$, thus resulting in a stronger field confinement.

The influence of the geometric parameters $\Lambda$ and $d/\Lambda$ has been accurately investigated through the FEM full-vector modal solver [2.11–2.13]. Five values of the hole-to-hole spacing $\Lambda$, that is 1, 1.5, 2, 2.5, and 3 μm, have been chosen, and $d/\Lambda$ has been varied in the range 0.5–0.9. In particular, PCFs with five rings of air-holes in the cross-section have been considered. It is important to underline that for fibers with low $\Lambda$ values, which have the highest leakage losses, the FEM solver with PML as boundary conditions has been used, not to affect the simulation results, as described in Appendix A.

Figure 2.2 shows the dispersion curve $n_{\text{eff}}(\lambda)$ of the square-lattice PCFs with different $d/\Lambda$ values and $\Lambda = 1$, 2, and 3 μm, respectively, for the wavelengths between 1200 and 1600 nm. As expected, for a fixed $\Lambda$ value, the effective index decreases in all the considered wavelength range as $d/\Lambda$ becomes higher, that is the air-filling fraction of the photonic crystal cladding increases. Moreover, as reported in Fig. 2.2a, when $\lambda$ changes from 1200 to 1600 nm, the $n_{\text{eff}}$ values become lower, and the slope of the dispersion curve increases with $d/\Lambda$. This is confirmed also in Fig. 2.2b and c for a pitch $\Lambda$ of 2 and 3 μm, respectively.

Looking at Fig. 2.3, it is possible to understand how the effective index of the square-lattice PCF fundamental mode changes as a function of the pitch

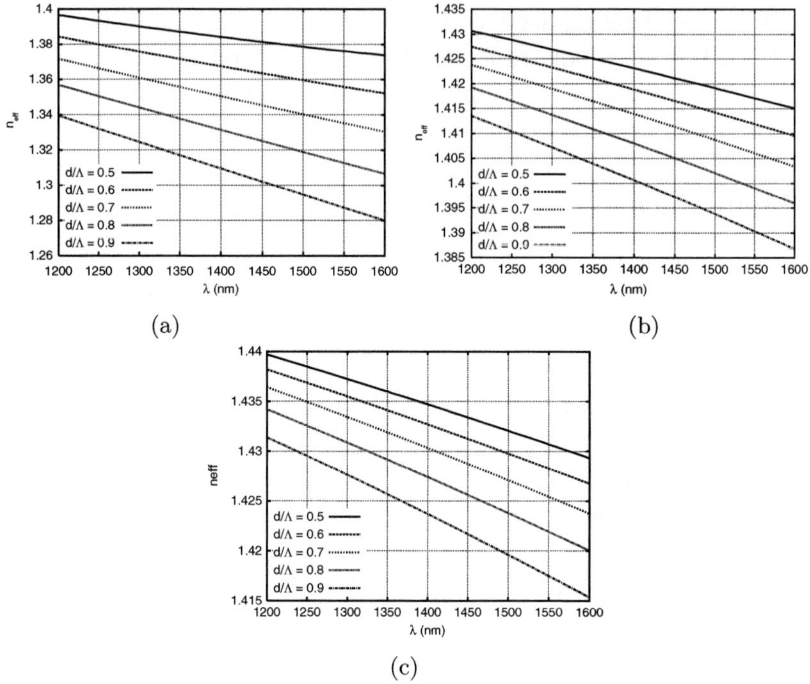


Figure **2.2**: The effective index $n_{\text{eff}}$ versus the wavelength of the square-lattice PCFs with (a) $\Lambda = 1$ μm, (b) $\Lambda = 2$ μm, and (c) $\Lambda = 3$ μm for different $d/\Lambda$ values in the range 0.5–0.9 [2.1].

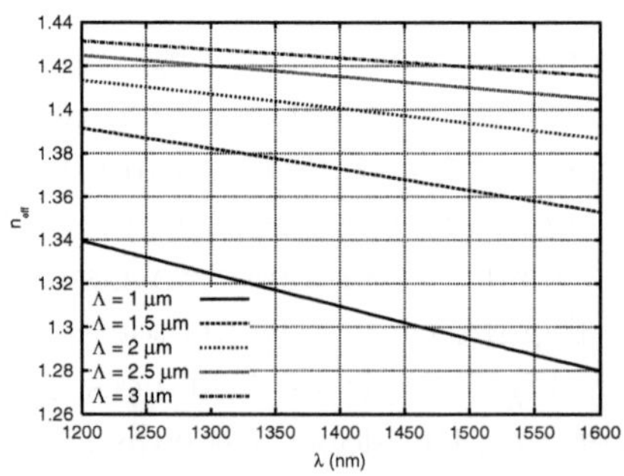


Figure **2.3**: The effective index $n_{\text{eff}}$ versus the wavelength of the square-lattice PCFs with $d/\Lambda = 0.9$ for different $\Lambda$ values between 1 and 3 μm [2.1].

$\Lambda$ for a fixed $d/\Lambda$ value. In this case $d/\Lambda = 0.9$ has been chosen, but results are almost the same for the other air-hole dimensions considered in the analysis. Notice that the highest $n_{\text{eff}}$ values have been obtained for the larger pitch, that is $\Lambda = 3$ μm. Moreover, the effective index decreases with the hole-to-hole spacing in all the considered wavelength range. In particular, the decrease of 0.5 μm in the pitch value, from 1.5 to 1 μm, causes the most significant change in the effective index, which is, for example, 1.358 and 1.287 at 1550 nm, respectively.

In order to make a comparison of the guiding properties of PCFs with different geometric characteristics, a square-lattice PCF and a triangular one have been considered with five air-hole rings and the same values of $\Lambda$ and $d/\Lambda$. A small $d/\Lambda$ value, that is 0.5, has been chosen for the two fibers, so that the triangular PCF is single-mode in all the wavelength range considered also for the largest pitch $\Lambda = 3$ μm [2.14, 2.15]. The dispersion curves $n_{\text{eff}}(\lambda)$ are reported in Fig. 2.4a for $\Lambda = 1$ μm and $\Lambda = 3$ μm. Notice that the fundamental mode of the square-lattice PCFs has a higher effective index value for both the considered pitch values. Moreover, the $n_{\text{eff}}$ difference between square and triangular PCFs with the same geometric parameters is higher for the smaller pitch, that is 1 μm.

A further comparison has been made between the square-lattice PCFs and the triangular ones, taking into account the effective area of the guided mode, evaluated according to Eq. (A.7). As shown in Fig. 2.4b for $d/\Lambda = 0.5$, the

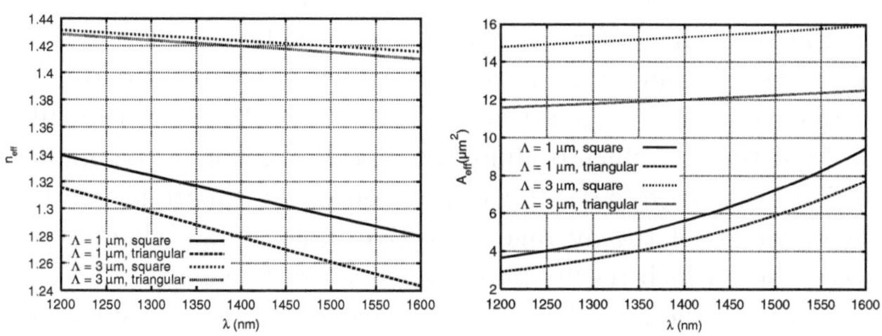


(a) (b)

Figure **2.4**: Comparison of (a) the effective index and (b) the effective area values for the square-lattice PCFs and the triangular ones with $d/\Lambda = 0.5$, for $\Lambda = 1$ and 3 μm [2.1].

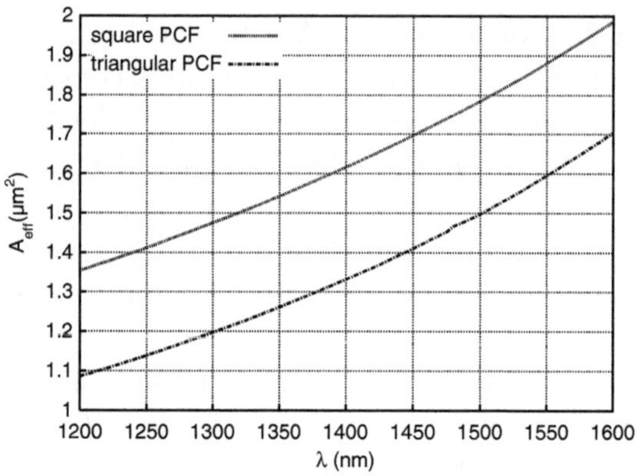


Figure **2.5**: Comparison of the effective area values for the square-lattice PCF and the triangular one with $d/\Lambda = 0.9$ and $\Lambda = 1$ μm [2.1].

PCFs with the square lattice have larger effective area for both the pitch values considered. In particular, there is a quite greater difference between the $A_{\text{eff}}$ values of the two kinds of PCFs if the pitch is large, that is $\Lambda = 3$ μm. The same behavior has been obtained for different geometric parameter values, that is $\Lambda = 1$ μm and $d/\Lambda = 0.9$, as reported in Fig. 2.5. It is important to underline that the effective area values of the square-lattice PCF are still small, being lower than 2 μm$^2$ in all the wavelength range considered, even if they are higher than those of the triangular PCF. As an example, the square-lattice PCF has an effective area at 1550 nm which is 18% larger than that of the fiber with the triangular lattice. This difference can be explained by considering the different air-hole position around the silica core, which is smaller for the triangular PCFs. Moreover, the square lattice is characterized by a lower air-filling fraction $f = (\pi/4)(d/\Lambda)^2$, which is almost 86% of the one for the triangular lattice, that is, $f = (\pi/2\sqrt{3})(d/\Lambda)^2$. As a consequence, the square-lattice PCFs provide higher values of the average refractive index of the cladding, that is a lower step index, which results in a lower field confinement.

Finally, notice the tight field confinement, due to the large core diameter, obtained in both the PCFs with $\Lambda = 3$ μm and $d/\Lambda = 0.5$, as shown in Fig. 2.6, and $d/\Lambda = 0.9$, as shown in Fig. 2.7, respectively. Since the fundamental component of the magnetic field is all confined inside the first air-hole ring, its shape clearly underlines the differences in the position of the air-holes belonging to the inner ring and, as a consequence, the different geometric

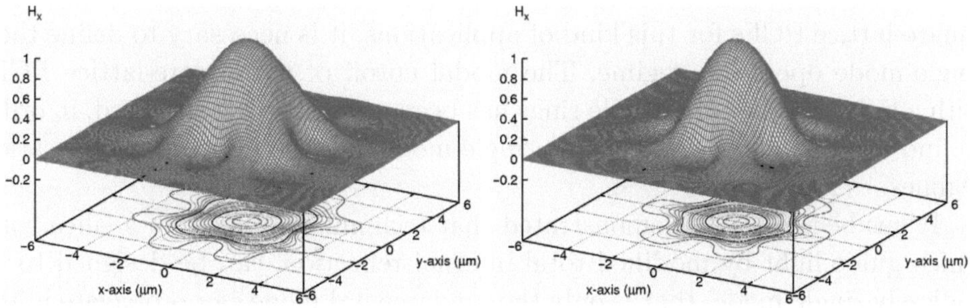


(a) (b)

Figure **2.6**: Fundamental component of the magnetic field at 1550 nm for (a) the square-lattice PCF and (b) the triangular one with $d/\Lambda = 0.5$ and $\Lambda = 3$ μm [2.1].

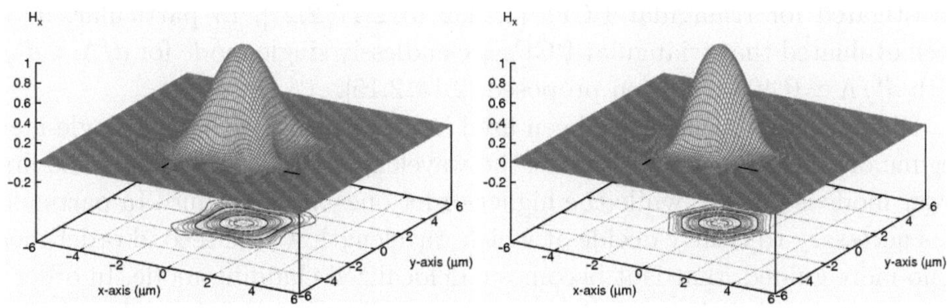


(a) (b)

Figure **2.7**: Fundamental component of the magnetic field at 1550 nm for (a) the square-lattice PCF and (b) the triangular one with $d/\Lambda = 0.9$ and $\Lambda = 3$ μm [2.1].

characteristics of the two lattices, that is the square and the triangular one. Notice that, due to their field shape, square-lattice PCFs could be useful if applied as pig-tail fibers of integrated optical devices with a rectangular or a square transverse section.

### 2.1.2 Cutoff

As it has been previously shown, square-lattice PCFs present a wider effective area than triangular ones for fixed $d/\Lambda$ and $\Lambda$ values, so they can be of practical interest as LMA fibers for high-power delivery. In order to successfully use square-lattice PCFs for this kind of applications, it is necessary to define their single-mode operation regime. The modal cutoff of the square-lattice PCFs with a finite number of air-hole rings has been accurately investigated, in order to find the boundary between the single-mode and the multi-mode operation regimes.

It has been already demonstrated that triangular PCFs with a silica core, which guide light by modified total internal reflection, can be designed to be endlessly single mode, that is only the fundamental mode can propagate in the fiber core for all the wavelengths, unlike conventional fibers which exhibit a cutoff wavelength below which higher-order modes are supported [2.16, 2.17]. A cutoff analysis for PCFs is not trivial as for conventional optical fibers because all the modes propagating in PCFs with a finite air-hole ring number are leaky [2.18–2.20]. The single-mode regime has been already successfully investigated for triangular PCFs [2.14, 2.15, 2.17, 2.21]. In particular, it has been evaluated that triangular PCFs are endlessly single mode for $d/\Lambda < d^*/\Lambda$ with $d^*/\Lambda \simeq 0.406$ has been proposed [2.14, 2.15].

Different approaches have been used in literature to study the single-mode regime of triangular PCFs, that is the wavelength range where only the first-order mode is guided, while the higher-order ones are unbound. In particular, it is necessary to clearly decide at which wavelength $\lambda^*$ the second-order mode is no more guided, that is it becomes a delocalized cladding mode. In order to find this transition, it is possible to take into account the divergence at long wavelengths of its effective area [2.22], or its leakage losses, which are related to the attenuation constant $\alpha$, the real part of the complex propagation constant in Eq. (A.2) [2.18, 2.20]. In particular, the normalized cutoff wavelength $\lambda^*/\Lambda$ can be evaluated by observing the transition shown by the behavior of $\alpha/k_0$, $k_0$ being the wave number, versus $\lambda/\Lambda$ [2.21]. This can be made evident by calculating the $Q$ parameter

$$Q = \frac{d^2 \log[\alpha/k_0]}{d^2 \log(\Lambda)} \,, \qquad (2.1)$$

because it exhibits a sharp negative minimum at $\lambda^*/\Lambda$ [2.21]. Here, the phase diagram with single-mode and multi-mode operation for square-lattice PCFs has been obtained by calculating the $Q$ parameter for different normalized wavelength $\lambda/\Lambda$ and by evaluating its negative minimum for PCFs with $d/\Lambda$ in the range 0.45–0.57. The analysis has been developed by fixing the guided-mode wavelength at 633 nm, as well as at 1550 nm. The hole-to-hole distance $\Lambda$

has been properly selected to obtain the desired normalized wavelength value. Due to the strong influence of the air-hole ring number on the leakage losses of PCFs with a finite cross-section [2.18, 2.19], fibers with four, six, and eight rings have been considered for the modal cutoff analysis. In fact, it has been already demonstrated that the transition of the $Q$ parameter becomes more acute and the method more reliable as the ring number increases [2.21]. Finally, it is important to point out the numerical methods used in this analysis. The complex propagation constants of the fundamental and the second-order mode, as well as the field distributions, have been calculated by means of the FEM full-vector modal solver with anisotropic PML [2.18, 2.20], as described in Appendix A. The multipole method [2.23, 2.24] has been also used to confirm the simulation results, obtaining a good agreement.

In order to calculate the $Q$ parameter according to Eq. (2.1), the behavior of $\alpha/k_0$ versus the normalized wavelength $\lambda/\Lambda$ for the second-order mode has been evaluated. As shown in Fig. 2.8a for eight-ring PCFs with different $d/\Lambda$ values, $\alpha/k_0$ increases with $\lambda/\Lambda$, that is the confinement of the second-order mode is lower for smaller pitch $\Lambda$.. For all the considered $d/\Lambda$ ratios the curves show a transition, that is a change in the slope, which becomes sharper as the air-hole diameter increases with respect to the pitch. Moreover, by varying $d/\Lambda$ from 0.45 to 0.57 the transition region moves toward the higher $\lambda/\Lambda$ values,

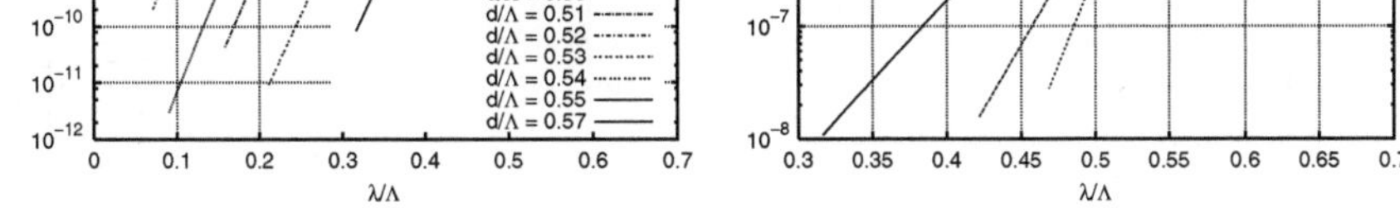


(a) (b)

Figure **2.8**: Second-order mode $\alpha/k_0$ versus the normalized wavelength $\lambda/\Lambda$ (a) for 8-ring square-lattice PCFs with $d/\Lambda$ in the range 0.45–0.57 and (b) as a function of the air-hole ring number, that is four, six, or eight, for a square-lattice PCF with $d/\Lambda = 0.57$ [2.2].

as it has been already demonstrated for triangular PCFs [2.21]. In addition, notice that, when $d/\Lambda = 0.45$, it is difficult to identify the transition, which, on the contrary, is very sharp when $d/\Lambda = 0.57$. The same behavior of $\alpha/k_0$ has been obtained for square-lattice PCFs with a lower air-hole ring number, that is four and six. However, it must be observed that, in these cases, as shown in Fig. 2.8b, the transition is not so sharp even for a high $d/\Lambda$ value.

From the previous results, the $Q$ parameter has been calculated through a finite difference formula and the values obtained for the eight-ring square-lattice PCFs are reported in Fig. 2.9a. The negative value of the curve minimum becomes higher as $d/\Lambda$ increases, reaching $-654$ at $\lambda/\Lambda \simeq 0.532$ for $d/\Lambda = 0.57$. As the square-lattice PCF air-filling fraction decreases, the $Q$ minimum moves toward the lower $\lambda/\Lambda$ values and becomes wide and difficult to identify with high precision. For example, the negative minimum almost disappears for the PCFs with $d/\Lambda = 0.45$, so its curve has not been drawn in the figure. A similar behavior has been obtained also for the PCFs with less air-hole rings. Fig. 2.9b, for example, reports data for the PCFs with $d/\Lambda = 0.57$, showing that the $Q$ minimum becomes less negative and moves toward higher $\lambda/\Lambda$ values when the ring number decreases. In particular, for four-ring fibers the dip is very wide and the most negative value is only $-73$ at $\lambda/\Lambda \simeq 0.571$, while it is $-260$ at $\lambda/\Lambda \simeq 0.541$ when the square-lattice PCFs have six air-hole rings.

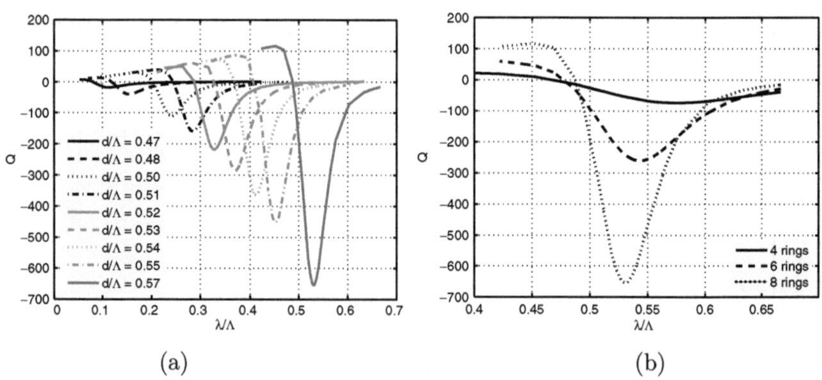


Figure **2.9**: $Q$ parameter values versus the normalized wavelength $\lambda/\Lambda$ (a) for 8-ring square-lattice PCFs with $d/\Lambda$ in the range 0.45–0.57 and (b) as a function of the air-hole ring number, that is, 4, 6, or 8, for a square-lattice PCF with $d/\Lambda = 0.57$ [2.2].

In summary, Figs. 2.8 and 2.9 clearly show that, when the leakage behavior is strong, whatever the reason, for example, low $d/\Lambda$ or few air-hole rings, it is difficult to define the transition region and the related cutoff wavelength. On the contrary, by considering a high number of air-hole rings the slope change in $\alpha/k_0$ is more evident, the $Q$ curve presents a sharp dip and it is possible to find reliable values of the normalized cutoff wavelength $\lambda^*/\Lambda$. These values for the square-lattice PCFs with eight air-hole rings are reported in Fig. 2.10, which also shows data for four-and six-ring PCFs. Notice that the $\lambda^*/\Lambda$ values have been reported only for the well-defined and sharp minima, that is for $d/\Lambda \geq 0.48$ for eight-ring PCFs and for $d/\Lambda \geq 0.50$ for four-, and six-ring PCFs. As expected, results change by increasing the air-hole ring number, tending to the values of a PCF with an ideal infinite cladding. This suggests again that the $Q$ parameter method must be applied assuming a high ring number.

This conclusion is confirmed also by further comments on the results reported in Fig. 2.10. In fact, it seems that PCFs with four air-hole rings have a smaller single-mode region, defined by $\lambda/\Lambda > \lambda^*/\Lambda$, their cutoff values being the highest ones. However, this result is in contradiction with the $\alpha/k_0$ values reported in Fig. 2.8b, which are also the highest for all the considered $\lambda/\Lambda$. Figure 2.8b, in fact, indicates that the second-order mode suffers from high leakage losses and consequently only the fundamental mode can actually propagate in a wider single-mode spectral range. In other words, the $Q$

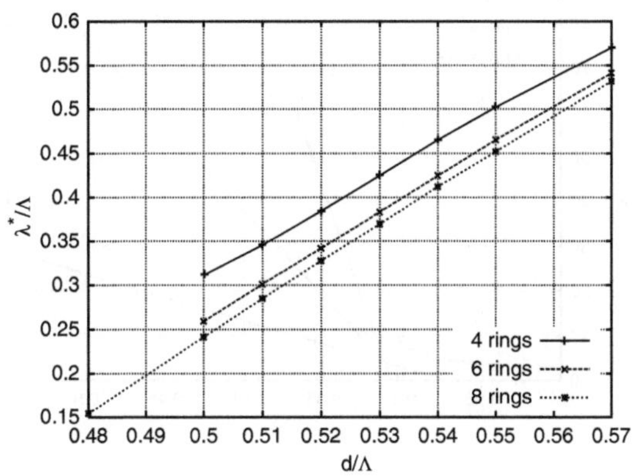


Figure **2.10**: Normalized cutoff wavelength $\lambda^*/\Lambda$ as a function of the $d/\Lambda$ ratio for square-lattice PCFs with four, six, and eight air-hole rings [2.2].

parameter approach fails when a sharp minimum does not occur, as shown in Fig. 2.9b for the case of four air-hole rings. On the contrary, by considering eight air-hole rings, results are clearly readable and reliable. It is important to highlight that the $\lambda^*/\Lambda$ evaluated for PCFs with many rings of air-holes also apply to fibers with few rings, being $\lambda^*/\Lambda$, in any case, an upper limit of the cutoff wavelength. This means that fibers with a reduced number of rings present an even larger single-mode region.

In order to give a further confirmation of what stated, the normalized cutoff wavelength has been evaluated also according to another approach, the method based on the second-order mode effective area proposed in [2.22]. Simulation results for the PCFs with $d/\Lambda = 0.52$ are shown in Fig. 2.11. Notice that the $\lambda^*/\Lambda$ values, indicated by the crossing of the solid lines with the horizontal axis, are, respectively, 0.273, 0.302, and 0.308 for the PCFs with four, six, and eight air-hole rings. This means that $\lambda^*/\Lambda$ increases with the air-hole ring number, that is the PCFs which provide the better field confinement have the smallest single-mode operation region and not the other way round, as could be suggested by Fig. 2.10. Moreover, the difference between the normalized cutoff wavelength values almost vanishes if PCFs with six, and eight rings are considered. Thus, eight-ring square-lattice PCFs offer the most reliable results and, in the following, will also be used to compare square and triangular lattice PCF characteristics.

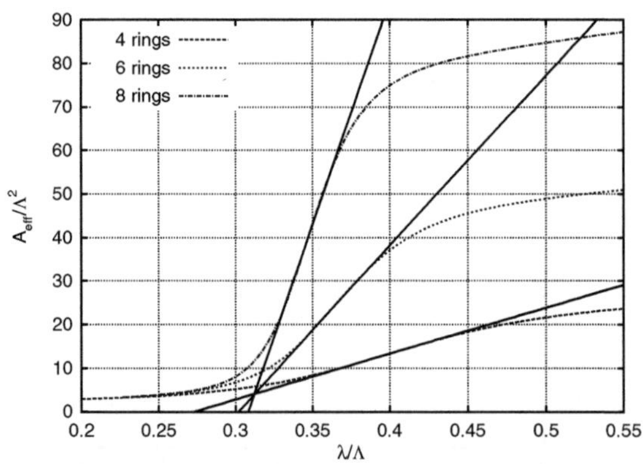


Figure **2.11**: Second-order mode normalized effective area $A_{\text{eff}}/\Lambda^2$ versus $\lambda/\Lambda$ for square-lattice PCFs with $d/\Lambda = 0.52$ and with four, six, and eight air-hole rings [2.2].

A first interesting comparison can be made on the endlessly single-mode region. For fibers with a triangular lattice of air-holes, a fitting of the cutoff curve has been evaluated according to the expression [2.21]:

$$\lambda^*/\Lambda \simeq \alpha \cdot (d/\Lambda - d^*/\Lambda)^\gamma \,, \tag{2.2}$$

where $d^*/\Lambda$ is the boundary of the endlessly single-mode region, resulting in $d^*/\Lambda = 0.406$, $\alpha = 2.80 \pm 0.12$, and $\gamma = 0.89 \pm 0.02$ [2.21]. The same procedure, applied to the $\lambda^*/\Lambda$ values of the square-lattice PCFs reported in Fig. 2.10, provides $d^*/\Lambda \simeq 0.442$, $\alpha = 4.192 \pm 0.246$ and $\gamma = 1.001 \pm 0.025$. The boundary between the single-mode and the multi-mode operation area is reported in Fig. 2.12 for square-lattice PCFs and triangular ones. Notice that the single-mode region for square-lattice PCFs, that is the one above the curve in Fig. 2.12, is wider for lower $d/\Lambda$ values, while the difference is significantly reduced until it disappears as the air-filling fraction increases. Moreover, it can be noticed that the $d^*/\Lambda$ value is higher for square-lattice fibers, that is they can be endlessly single mode in a wider range of the geometric parameter values with respect to triangular PCFs, and they can be successfully used in applications which need large-mode area fibers.

As a second part of the cutoff analysis, starting from the single-mode regime information obtained with the $Q$ parameter approach, the normalized cutoff frequency $V^*$ has been evaluated. The $V$ parameter can be easily calculated in a standard optical fiber, since it depends on the core radius and

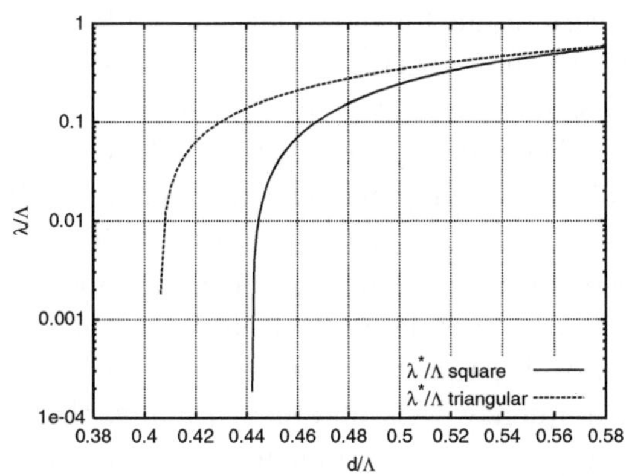


Figure **2.12**: Phase diagram for eight air-hole ring PCFs characterized by the square and the triangular lattice [2.2].

the core and cladding refractive indices, which are all well defined. The choice of these parameters for PCFs is not trivial, and several formulations of the normalized frequency have been proposed in literature [2.15, 2.17, 2.25–2.28], based either on geometric and physical considerations, or analogies with classical theory of conventional fibers. In this study two different formulations of the $V$ parameter are considered. The first one is

$$V_1 = \frac{2\pi}{\lambda}\Lambda\sqrt{n_{\rm eff}^2 - n_{\rm FSM}^2}\,, \tag{2.3}$$

which has been recently proposed for triangular PCFs [2.15, 2.29]. In Eq. (2.3) $n_{\rm eff}$ and $n_{\rm FSM}$ are the effective indices, respectively, of the fundamental guided mode and of the fundamental space-filling mode in the air-hole cladding, which has been evaluated using a freely available software package [2.30]. The choice of $\Lambda$ as the effective core radius can be adopted also for the PCFs here studied, since it is the natural length scale of both the triangular and the square lattices [2.15, 2.29]. The second $V$ parameter definition considered, more similar to the one used for conventional fibers, is

$$V_2 = \frac{2\pi}{\lambda}\rho\sqrt{n_{\rm co}^2 - n_{\rm FSM}^2}\,, \tag{2.4}$$

where $n_{\rm co}$ is the refractive index of the silica core at the operation wavelength, and $\rho$ is the effective core radius. In order to properly adapt the concept of the $V$ parameter to PCFs, several values for $\rho$ have been proposed in literature for fibers characterized by a triangular lattice, that is $0.5\Lambda$ [2.31], $\Lambda/\sqrt{3}$ [2.27, 2.28], $0.625\Lambda$ [2.25], $0.64\Lambda$ [2.26], and $\Lambda$ [2.17, 2.25]. In the present study the effective core radius for the square-lattice PCFs has been considered equal to $0.67\Lambda$. This value, different from all the others previously adopted for triangular PCFs, has been evaluated through the method proposed by Brechet et al. [2.26]. The technique consists in calculating the refractive index of the fundamental space-filling mode $n_{\rm FSM}$ and assessing a temporary $V$ parameter $V_t$ according to Eq. (2.4) with $\rho = \Lambda$. Then, using the effective index of the guided mode $n_{\rm eff}$, the normalized propagation constant $\beta_n = (n_{\rm eff}^2 - n_{\rm FSM}^2)/(n_{\rm co}^2 - n_{\rm FSM}^2)$ is determined. Substituting the $\beta_n$ value into the characteristic equation for the step-index fibers with $NA = (n_{\rm co}^2 - n_{\rm FSM}^2)^{1/2}$, a new normalized frequency $V$ is obtained. Finally, the effective core radius is given by the ratio $\rho = V/V_t$. By plotting $\rho$ versus the normalized air-hole diameter $d/\Lambda$, it can be shown that, in the limit of short wavelengths compared to the air-hole size, that is $d/\lambda \geq 2$, and for low air-filling fractions, that is $d/\Lambda \leq 0.5$, the effective core radius tends to a constant value regardless of $d/\Lambda$.

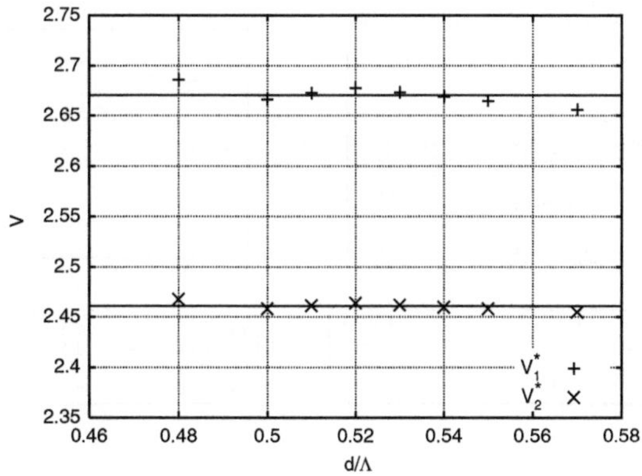


Figure **2.13**: Cutoff value $V^*$ of the normalized frequency according to the two definition for square-lattice PCFs with eight rings. Solid lines represent the mean value of $V_1^*$ and $V_2^*$ [2.2].

As shown in Fig. 2.13, $V_1^*$ and $V_2^*$ have been evaluated for the eight air-hole ring PCFs starting from the normalized cutoff wavelength at the $d/\Lambda$ values reported in Fig. 2.10. The mean values of $V_1^*$ and $V_2^*$, respectively 2.67 and 2.46, are also reported as a solid line in Fig. 2.13 and have been assumed as reference values like 2.405 for a standard fiber. Figure 2.14a and b show the $V$ number versus the normalized wavelength calculated according to Eqs. (2.3) and (2.4), and the corresponding $V^*$ mean value as a horizontal solid line. Of course the crossings between the $V^*$ line and the $V$ number curves for the two formulations give again the $\lambda^*/\Lambda$ behavior versus $d/\Lambda$, that is the single-mode–multi-mode phase diagram of Fig. 2.10.

Finally, it is important to notice that the value of $V_1^*$ here evaluated for the square-lattice PCFs is lower than $\pi$, the value for the triangular PCFs [2.15], which has been obtained with the same $V$ number expression and by looking at the second-order mode field distribution on the fiber cross-section [2.15, 2.29]. In particular, it has been shown that in triangular PCFs the second-order mode effective transverse wavelength, related to the dimension of the defect region where the mode fits in, is $\lambda_\perp^* \simeq 2\Lambda$ at the cutoff condition. As a consequence, the normalized cutoff frequency becomes $V_1^* = \dfrac{2\pi}{\lambda_\perp^*}\Lambda \simeq \pi$ [2.15]. In order to extend the same approach to the square-lattice PCFs, the magnetic field components shown in Fig. 2.15 have to be taken into account. It is important

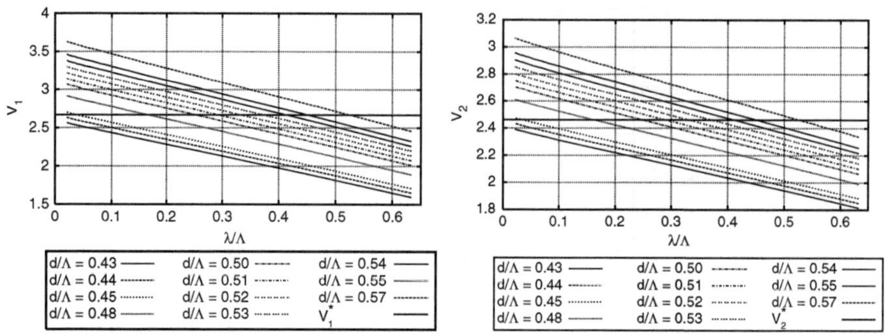


(a) (b)

Figure **2.14**: (a) $V_1$ and (b) $V_2$ behavior versus the normalized wavelength $\lambda/\Lambda$ for square-lattice PCFs with $d/\Lambda$ between 0.43 and 0.57. A solid horizontal line is drawn at the fixed value $V_1^*$ and $V_2^*$, respectively [2.2].

to underline that the field shape of the second-order mode in these PCFs is strongly influenced by the fourfold symmetry which characterizes the square lattice, in particular by the position of the air-holes belonging to the first ring. As a consequence, different $\lambda_\perp^*$ values can be obtained if the second-order mode field amplitude is considered along the horizontal, or vertical, direction, or along the 45° one. The two situations are depicted in Fig. 2.16a and b. In the first case, the field shape is the same of the one reported for the triangular PCFs [2.15], so $\lambda_\perp^* \simeq 2\Lambda$ and $V_1^* \simeq \pi$. On the contrary, if the 45° direction is considered, the separation between the two first null values of the second-order mode field amplitude increases, as shown in Fig. 2.16b, since the two opposite air-holes belonging to the first ring are more distant. Thus $\lambda_\perp^*$ is higher, that is $2\sqrt{2}\Lambda$, and consequently $V_1^* \simeq \dfrac{\pi}{\sqrt{2}}$. It is interesting to point out that the $V_1^*$ value calculated in the present analysis, that is 2.67, is almost equal to the mean value between $\pi$ and $\dfrac{\pi}{\sqrt{2}}$, that is 2.68. The corresponding $\lambda_\perp^* \simeq 2.34\Lambda$ is obtained by the mean value of the inverse of $2\Lambda$ and $2\sqrt{2}\Lambda$. In conclusion, it is not possible to simply extend the derivation of $V_1^*$ previously proposed for triangular PCFs to the case of square-lattice PCFs, since a unique value of $\lambda_\perp^*$ can not be easily found.

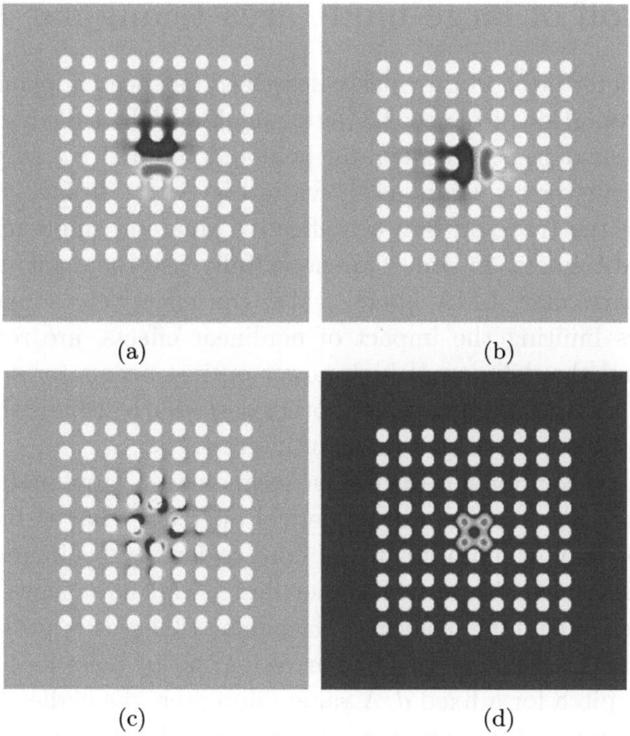


Figure **2.15**: (a) $H_x$, (b) $H_y$, (c) $H_z$, and (d) intensity distribution of the second-order guided mode at $\lambda/\Lambda \simeq 0.127$ for a four-ring square-lattice PCF with $d/\Lambda = 0.57$ [2.2].

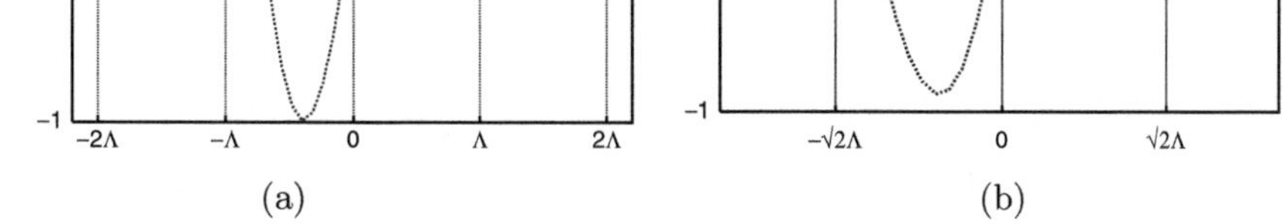


Figure **2.16**: Section of the square-lattice PCF cross-section (*solid line*) and of the $H_x$ field component (*dotted line*) (a) along the $x$-axis and (b) along the 45° direction [2.2].

## 2.2 Cutoff of large-mode area triangular PCFs

The $Q$ parameter method previously described has been applied also to study the cutoff properties of a new LMA triangular PCF, called seven-rod core, obtained by removing the central air-hole and the first six surrounding ones in the fiber transverse section, as shown in Fig. 2.17b. In fact, it is important to investigate the trade-off between effective area and single-mode operation regime in LMA fibers, in order to successfully use them for different applications. In particular, LMA fibers, which can effectively support high optical intensities limiting the impact of nonlinear effects, are required for the generation and the delivery of high-power optical beams for a wide range of applications. For such applications another desirable feature is the single-mode operation over the wavelength range of interest.

Using the conventional optical fiber technology, a large modal area can be achieved either by reducing the numerical aperture, that is by lowering the percentage of doping material in the core region, or by increasing the core dimension. Better results in LMA fiber design can be reached by exploiting PCFs. In particular, by considering triangular PCFs, it is possible to significantly increase the effective area by narrowing the air-holes for a fixed $\Lambda$, or by enlarging the pitch for a fixed $d/\Lambda$ value. Moreover, the endlessly single-mode property can provide the single-mode operation [2.32]. However, an upper limit on the guided-mode area exists, given by the value of the losses. In fact, the air-filling fraction decrease can cause an increase of the leakage losses [2.18, 2.19],

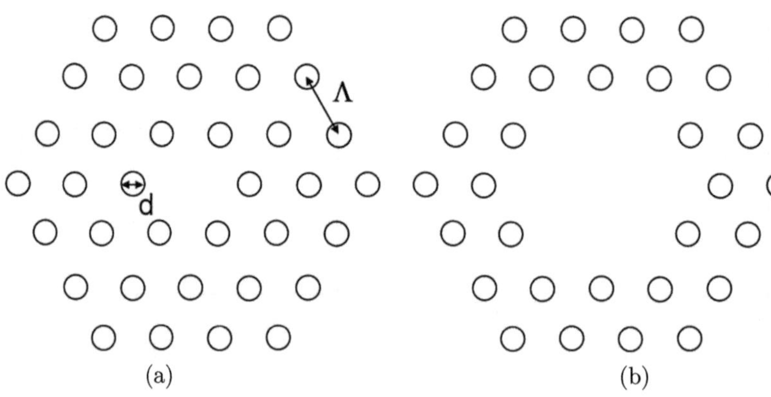


Figure **2.17**: (a) One-rod and (b) seven-rod core triangular PCF cross-section [2.3].

while, as $\Lambda$ becomes larger, there is a greater susceptibility to scattering losses induced by microbending and macrobending [2.33].

Another LMA PCF design based on the triangular lattice has been proposed in [2.33,2.34]. The triangular core region of these fibers, called three-rod core triangular PCFs, has been obtained by removing three air-holes in the center of the fiber cross-section. Three-rod core PCFs can provide an enhancement of the guided-mode area of about 30% and a higher robustness when scaled to a larger pitch [2.33]. As a drawback of the larger silica core dimension, the ESM region of these PCFs is smaller than that of the traditional triangular fibers, being limited by $d/\Lambda < 0.25$. Moreover, the triangular core symmetry influences the shape of the guided-mode field intensity, which deviates from the standard gaussian-like one.

In order to overcome these problems, a new triangular PCF can be considered instead of the traditional one-rod core fiber, reported in Fig. 2.17a. It is characterized by a triangular lattice and a silica core formed by removing seven central air-holes, as shown in Fig. 2.17b, so it will be referred as seven-rod core PCF in the following. By removing the air-holes belonging to the first ring, a wider silica region has been obtained, so seven-rod core PCFs present a larger effective area for fixed $d/\Lambda$ and $\Lambda$ values, compared to one-rod core fibers. The structure here studied has been chosen so that it can be readily fabricated. In fact, the proposed geometry is feasible using the well-known stack-and-draw technique without any additional difficulty.

Since the core dimension has a strong influence on the confinement of all the PCF-guided modes and, as a consequence, on the single-mode regime of triangular fibers, it is necessary to accurately define the single-mode operation regime of these LMA triangular PCFs, in order to successfully use them for practical applications. To this aim, a detailed analysis of the seven-rod core PCF cutoff properties has been carried out with the method previously described, that is the $Q$ parameter method, based on the leaky nature of the second-order mode. In this case the negative minima of the $Q$ parameter have been evaluated for PCFs with $d/\Lambda$ in the range 0.08–0.32. 10 air-hole ring one-rod core triangular PCFs have been already used for the modal cutoff analysis [2.21]. In the present study seven-rod core PCFs with nine rings have been considered. In Fig. 2.18a, b, and c the second-order mode magnetic field transverse components distribution at $\lambda/\Lambda \simeq 0.369$ is reported for a LMA PCF with $d/\Lambda = 0.2$. At this normalized wavelength the first higher-order mode results confined in the fiber silica core. This is confirmed by the second-order mode intensity distribution, shown in Fig. 2.18d.

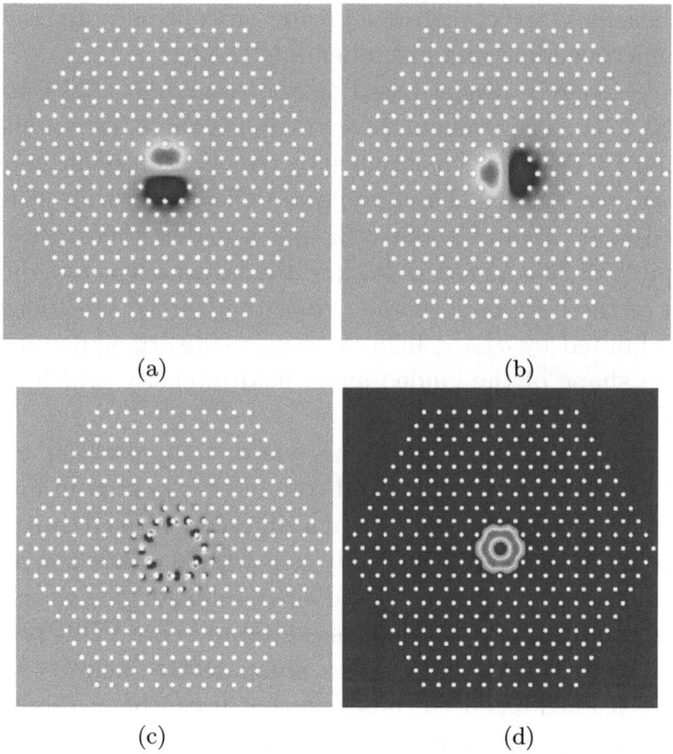


Figure **2.18**: (a) $H_x$, (b) $H_y$, (c) $H_z$, and (d) field intensity distribution of the second-order guided mode at $\lambda = 1550$ nm for the seven-rod core PCF with $d/\Lambda = 0.2$ and $\Lambda = 4.2$ μm [2.3].

The behavior of $\alpha/k_0$ versus the normalized wavelength $\lambda/\Lambda$ for the second-order mode, which has been calculated for seven-rod core PCFs with different $d/\Lambda$ values, is shown in Fig. 2.19a. Conclusions analogous to those reported in Section 2.1.2 for square-lattice PCFs can be drawn about the influence of the PCF geometric parameters, that is the pitch $\Lambda$ and the air-filling fraction $d/\Lambda$, on the $\alpha/k_0$ curves for seven-rod core triangular fibers. The $Q$ parameter has been then evaluated through a finite difference formula, and the values obtained are reported in Fig. 2.19b. Notice that the negative curve minimum becomes higher when $d/\Lambda$ increases, reaching $-270$ at $\lambda/\Lambda \simeq 1.19$ for $d/\Lambda = 0.32$. On the contrary, when the PCF air-filling fraction decreases, the $Q$ minimum shifts toward the lower $\lambda/\Lambda$ values, becoming wide and difficult

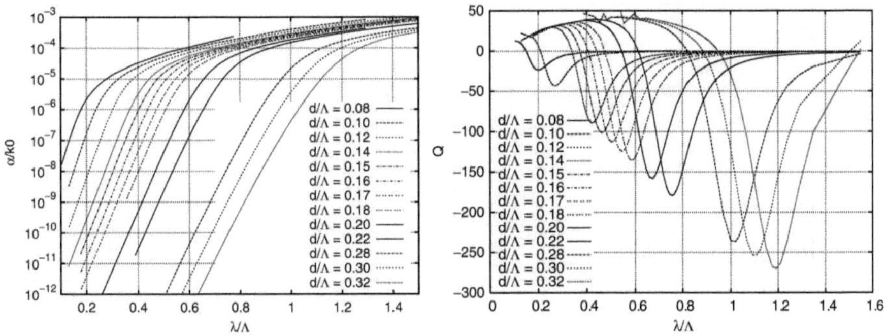


(a) (b)

Figure **2.19**: (a) Second-order mode $\alpha/k_0$ and (b) $Q$ parameter values as a function of the normalized wavelength $\lambda/\Lambda$ for nine-ring seven-rod core triangular PCFs with $d/\Lambda$ in the range 0.08–0.32 [2.3].

to identify with high precision. For example, the negative minimum is about $-23.5$ at $\lambda/\Lambda \simeq 0.2$ for the PCFs with $d/\Lambda = 0.08$.

A second approach has been applied to confirm the results obtained with the $Q$ method. In fact, the limit of the single-mode region can be determined by comparing the effective index $n_{\text{eff}} = \beta/k_0$ of the second-order mode and that of the fundamental space-filling mode $n_{\text{FSM}}$ for a fixed $d/\Lambda$ value [2.33, 2.34]. The first higher-order mode at a certain wavelength $\lambda$ is no longer guided if its $n_{\text{eff}}$ is lower than the $n_{\text{FSM}}$ at the same $\lambda$. As a consequence, the normalized cutoff wavelength $\lambda^*/\Lambda$ is obtained applying the condition $n_{\text{eff}} = n_{\text{FSM}}$.

The second-order mode effective index and the $n_{\text{FSM}}$ have been reported as a function of the normalized wavelength $\lambda/\Lambda$ in Fig. 2.20a for the seven-rod core triangular PCFs with $d/\Lambda = 0.2$ and $d/\Lambda = 0.28$. Notice that the value of the normalized cutoff wavelength, evaluated by considering the crossing of the $n_{\text{eff}}$ and $n_{\text{FSM}}$ curves, becomes higher as the air-filling fraction of the photonic crystal cladding increases, being 0.97 and 0.63 for $d/\Lambda$ equal to 0.28 and 0.2, respectively.

The $\lambda^*/\Lambda$ values calculated with both the previous methods for seven-rod core triangular PCFs with nine air-hole rings and $d/\Lambda$ in the range 0.08–0.32 are reported in Fig. 2.20b. Notice that the results obtained are in good agreement, even if the $Q$ parameter method is less precise for the PCFs with the lower air-filling fraction, being the evaluated minima wider and less deep for $d/\Lambda \leq 0.12$.

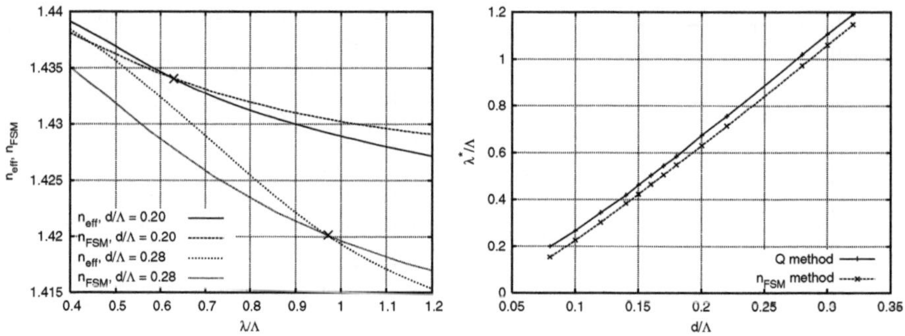


(a) (b)

Figure **2.20**: (a) Second-order mode $n_{\text{eff}}$ and $n_{\text{FSM}}$ versus the normalized wavelength $\lambda/\Lambda$ for nine-ring seven-rod core triangular PCFs with $d/\Lambda$ equal to 0.2 and 0.28. (b) Normalized cutoff wavelength $\lambda/\Lambda$ as a function of the $d/\Lambda$ ratio for seven-rod core triangular PCFs, obtained with the $Q$ parameter approach and the $n_{\text{FSM}}$ method [2.3].

An interesting comparison can be made on the ESM region of triangular PCFs with core defect regions of different dimension, obtained by removing one or seven air-holes in the cross-section center. The fitting reported in Eq. (2.2) has been applied to the $\lambda^*/\Lambda$ values of the seven-rod core PCFs obtained with the $Q$ method and reported in Fig. 2.20b, providing $d^*/\Lambda \simeq 0.035$, $\alpha = 4.432 \pm 0.067$ and $\gamma = 1.045 \pm 0.01$. The boundary between the single-mode and the multi-mode operation area for small- and large-core triangular PCFs is reported in Fig. 2.21. Notice that the single-mode region for seven-rod core PCFs, that is the one above the continuous line, is significantly smaller than that of one-rod core fibers for lower $d/\Lambda$ values, while the difference between the two cutoff curves is reduced as the air-filling fraction increases. Moreover, it is important to underline that seven-rod core PCFs are characterized by a lower $d^*/\Lambda$ value, that is they can be ESM in a smaller range of the geometric parameter values with respect to one-rod core triangular fibers. In particular, the LMA fibers here proposed are endlessly single-mode only for $d/\Lambda < 0.035$.

In order to give a complete description of the seven-rod core triangular PCF cutoff properties, the normalized cutoff frequency $V^*$ has been evaluated, starting from the single-mode regime information obtained with the two previous approaches. The formulation of the $V$ parameter in Eq. (2.4) has been considered. Notice that the effective core radius for seven-rod core triangular

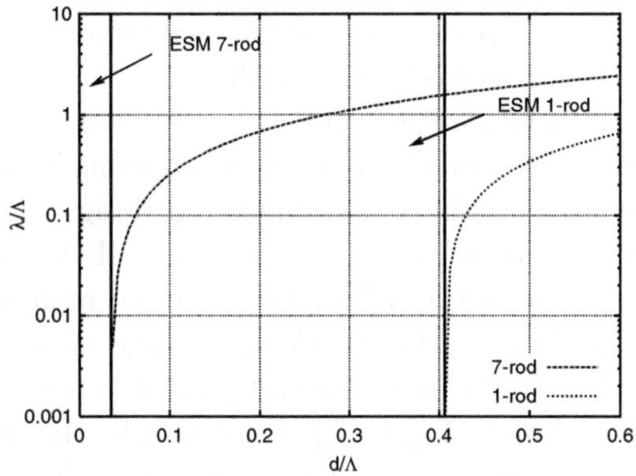


Figure **2.21**: Phase diagram for triangular PCFs with seven-rod, and one-rod core [2.3].

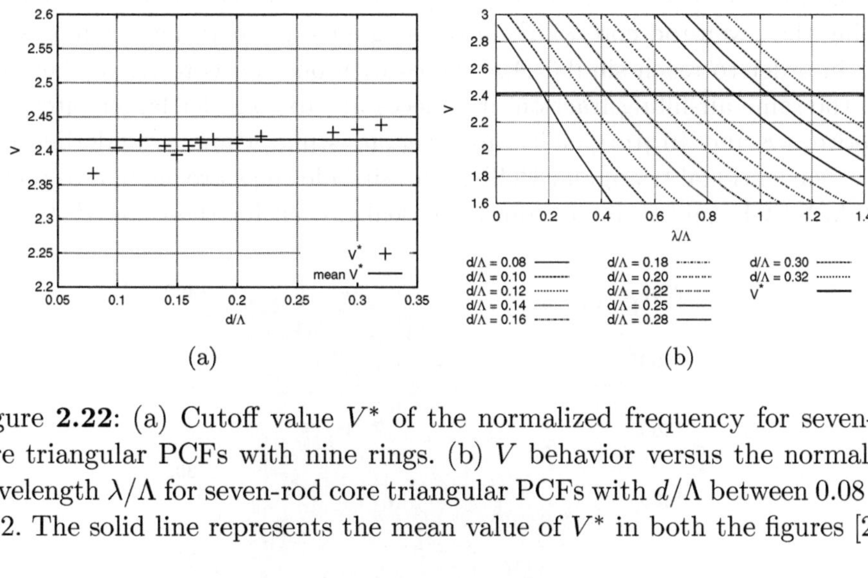


Figure **2.22**: (a) Cutoff value $V^*$ of the normalized frequency for seven-rod core triangular PCFs with nine rings. (b) $V$ behavior versus the normalized wavelength $\lambda/\Lambda$ for seven-rod core triangular PCFs with $d/\Lambda$ between 0.08 and 0.32. The solid line represents the mean value of $V^*$ in both the figures [2.3].

PCFs, evaluated through the method proposed by Brechet et al. [2.26] previously described, has been considered equal to $1.48\Lambda$. As shown in Fig. 2.22a, $V^*$ has been evaluated for the nine air-hole ring seven-rod core triangular PCFs, starting from the normalized cutoff wavelength at the $d/\Lambda$ values reported in Fig. 2.20b. The mean value of $V^*$, that is 2.416, is also shown as a solid line in Fig. 2.22a. Moreover, it is reported also as a horizontal solid line in Fig. 2.22b, where the $V$ parameter values, evaluated according to Eq. (2.4), are shown versus the normalized wavelength.

Since it has been demonstrated a strong correlation between the achievable guided-mode effective area and the single-mode regime, it becomes challenging to fulfill simultaneously all the requirements to design LMA seven-rod core triangular PCFs useful for practical applications. However, it is possible to find a compromise between the achievable effective area and the number of modes that PCFs guide over the wavelength range of interest. To this aim, the effective area $A_{\text{eff}}$ of the fundamental guided mode of LMA PCFs has been calculated, according to Eq. (A.7). For example, the $A_{\text{eff}}$ values obtained for seven-rod core PCFs with $\Lambda = 5.8$ μm, and nine air-hole rings are shown in Fig. 2.23 in the wavelength range 1000–2000 nm, as well as the boundary between the single-mode and the multi-mode region previously evaluated. In particular, it is possible to obtain an effective area at 1550 nm of about 320 and 268 $\mu m^2$, respectively, by choosing $d/\Lambda$ equal to 0.08 and 0.1, while still keeping the seven-rod core PCFs in the single-mode operation regime. Notice that, in order to reach similar $A_{\text{eff}}$ values with one-rod core triangular PCFs with the same air-filling fraction, it is necessary to consider larger pitch, that is between 8 and 10 μm [2.32]. Moreover, it is important to underline that, unlike conventional triangular PCFs, nine air-hole rings are enough to prevent the LMA PCF guided-mode from being leaky, even for these low $d/\Lambda$ and $\Lambda$

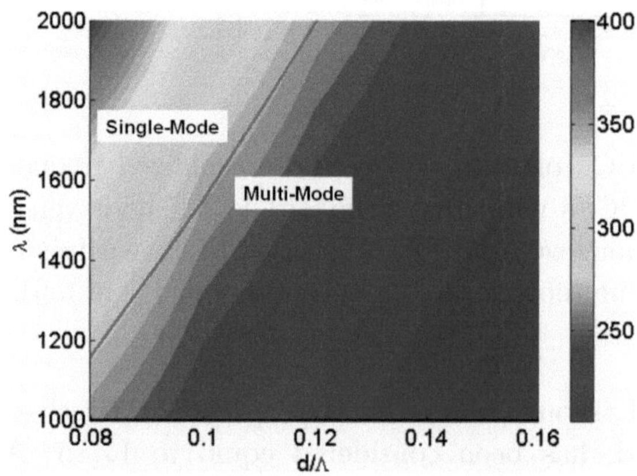


Figure **2.23**: The effective area $A_{\text{eff}}$ in $\mu m^2$ as a function of the air-filling fraction $d/\Lambda$ and the wavelength $\lambda$ for seven-rod core triangular PCFs with $\Lambda = 5.8$ μm, and nine air-hole rings [2.3].

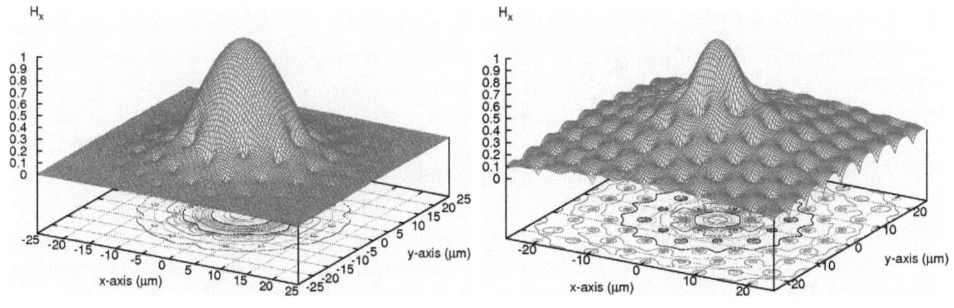


(a) (b)

Figure **2.24**: Magnetic field fundamental component at $\lambda/\Lambda = 0.267$ of (a) seven-rod and (b) one-rod core triangular PCFs with $d/\Lambda = 0.1$ [2.3].

values. In fact, the fundamental mode of the seven-rod core triangular PCF with $d/\Lambda = 0.1$ and $\Lambda = 5.8$ μm is completely confined in the silica core, as shown in Fig. 2.24a. On the contrary, if a one-rod core triangular PCF with the same geometric parameters and air-hole ring number, that is 10, is considered, the guided mode at 1550 nm is leaky, as it is shown in Fig. 2.24b. The small core dimension and the low air-filling fraction do not provide the necessary field confinement.

Finally, a further solution with an enlarged core region for a fixed $d/\Lambda$ and $\Lambda$ has been adopted, which should give a larger mode size then one-rod core triangular PCFs, without significantly increasing the guided-mode leakage losses. In particular, triangular PCFs with a silica core larger than that of one-rod core fibers, but smaller than that of seven-rod core ones have been considered. In fact, as represented in Fig. 2.25a, the diameter of the air-holes belonging to the first ring is $d_1 = 0.5d$ in the studied PCFs, while the air-hole ring number is still 10. Preliminary results of the cutoff analysis for these LMA triangular PCFs have been obtained and they are here reported. Looking at the magnetic field of the fundamental mode at 1550 nm guided by this kind of PCFs, shown in Fig. 2.25b, it is possible to notice a higher confinement with respect to one-rod core PCFs, even if the behavior of the guided mode is still leaky. Moreover, $\alpha/k_0$ curves for the second-order mode, evaluated as previously described for $d/\Lambda$ in the range 0.2–0.4 and reported in Fig. 2.26, do not present a net transition which describes the boundary

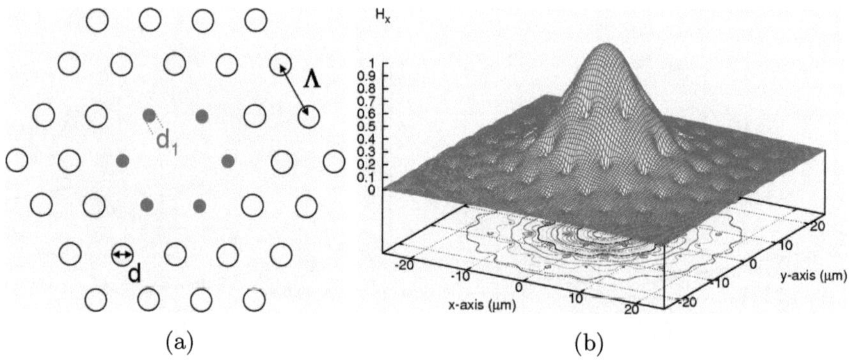


Figure **2.25**: (a) cross-section of the triangular PCF with $d_1 = 0.5d$. (b) Magnetic field fundamental component at $\lambda/\Lambda = 0.267$ of the triangular PCF with $d/\Lambda = 0.1$ and $d_1 = 0.5d$ [2.3].

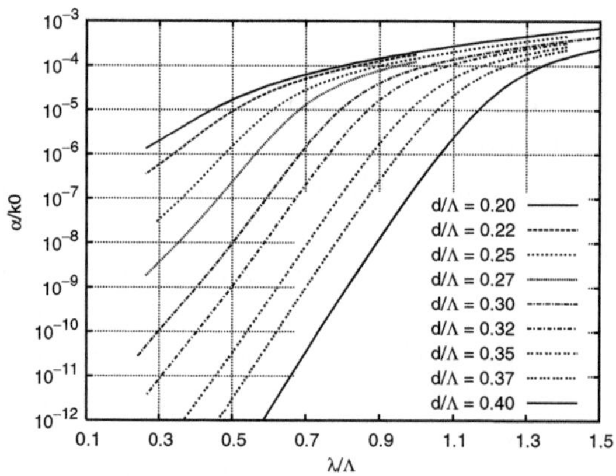


Figure **2.26**: Second-order mode $\alpha/k_0$ as a function of the normalized wavelength $\lambda/\Lambda$ for triangular PCFs with $d_1 = 0.5d$ for $d/\Lambda$ between 0.2 and 0.4 [2.3].

between the single-mode and the multi-mode region, differently from those reported in Fig. 2.19a for seven-rod core PCFs. As a consequence, more than 10 air-hole rings should be considered in order to successfully analyze the cutoff properties of these PCFs.

## 2.3 Hollow-core-modified honeycomb PCFs

While all the previous results regard the guiding properties of PCFs with a silica core, which guide light for TIR, here the analysis of the dispersion, the leakage, and the birefringence properties of hollow-core fibers which exploit the PBG effect, called photonic bandgap fibers (PBGFs), is reported. In particular, air-guiding has been studied in hollow-core PBGFs with a modified honeycomb air-hole lattice. Moreover, the influence of the hollow-core dimension, as well as of the cladding geometric parameters on the confinement loss and the single-mode behavior of the fibers has been investigated.

### 2.3.1 Guidance and leakage

The guiding properties of the hollow-core PCFs are mainly influenced by the wideness of the PBG crossed by the air-line. In particular, in a narrow PBG the light guidance is possible in a restricted wavelength range, thus causing high confinement loss. Different air-hole arrangements have been recently analyzed, in order to find an air-hole lattice with a wider PBG across the air-line. Among the different air-hole arrangements, PCFs with a triangular lattice have been usually employed for air-guiding [2.35, 2.36]. Unfortunately, in this case the PBG crossed by the air-line is quite narrow. Improvements have been reached by considering a triangular lattice with a high air-filling fraction, that is with large air-holes. This choice provides a wide PBG and, consequently, a better air guidance. Air-guiding in PCFs with honeycomb lattice has been also numerically demonstrated [2.37] and the leakage losses have been calculated [2.38, 2.39].

In the present analysis hollow-core PCFs with a modified honeycomb lattice, which has been proposed in [2.40], have been considered. In fact, it has been demonstrated that a wide bandgap crossed by the air-line can be obtained with this lattice, by properly choosing the geometric parameter values. All these aspects have been investigated in [2.40]. Figure 2.27 shows the unit cell of the modified honeycomb lattice. With respect to the original honeycomb geometry, shown on the left of the same figure, with hole-to-hole spacing $\Lambda$ and air-hole diameter $d$, an extra air-hole with diameter $d_c$ is added in the center of each cell. Notice that, when $d_c = 0$, the lattice degenerates into the basic honeycomb structure, whereas, when $d_c = d$, the lattice corresponds to

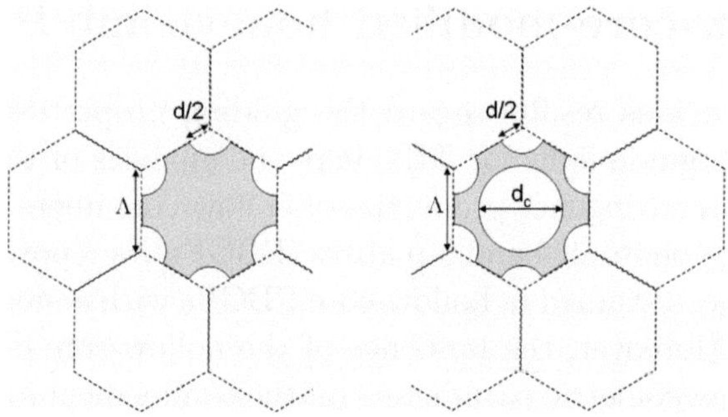


Figure **2.27**: (*Left*) Honeycomb unit cell. (*Right*) Modified honeycomb unit cell. Grey regions represent silica [2.7].

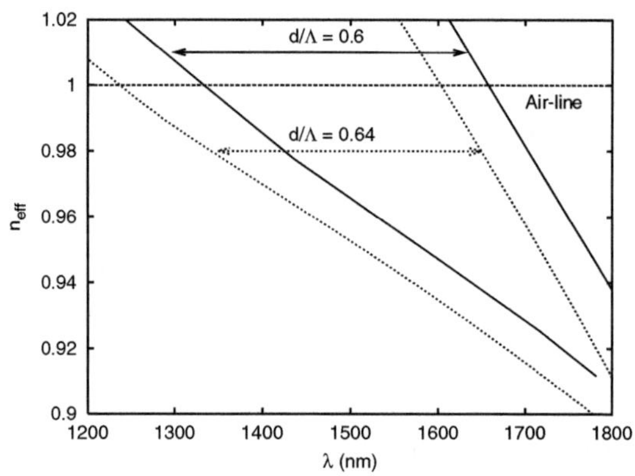


Figure **2.28**: Photonic bandgap edges for $d/\Lambda = 0.6$ (*solid line*) and $d/\Lambda = 0.64$ (*dotted line*) [2.7].

the triangular one. The air-filling fraction of the modified honeycomb lattice is given by

$$f = \frac{\pi}{3\sqrt{3}} \left[ \left( \frac{d}{\Lambda} \right)^2 + \frac{1}{2} \left( \frac{d_c}{\Lambda} \right)^2 \right] . \tag{2.5}$$

Moreover, the extra air-hole provides an additional degree of freedom in tailoring the PBG. Two different $d/\Lambda$ values have been chosen for the cladding, that is 0.6 and 0.64, while keeping fixed both $d_c/\Lambda = 1.32$ and $\Lambda = 1.62$ μm. The PBG edges calculated for these $d/\Lambda$ values are reported in Fig. 2.28. Notice that, when $d/\Lambda = 0.6$, which corresponds to an air-filling fraction $f = 74.4\%$, the air-line crosses the PBG from $\lambda = 1333$ to 1663 nm, that is in a wavelength range twice wider than that of a triangular lattice with the same air-filling fraction [2.41]. By increasing the air-filling fraction up to $d/\Lambda = 0.64$, the bandgap slightly enlarges, while shifting toward shorter wavelengths, so that the crossing with the air-line occurs at $\lambda = 1237$ and 1603 nm.

The hollow core of the PBGFs studied in the present analysis has been obtained by removing the silica inside a circle of radius $R$, as shown in Fig. 2.29. Four different fibers, A, B, C, and D, have been analyzed. In particular, fibers A and B, which have both a cladding with $d/\Lambda = 0.6$, are characterized by $R = 2\Lambda$ and $R = 3\Lambda$, respectively. Fibers C and D have the same hollow-core dimension of fibers A and B, respectively, but a different $d/\Lambda$ value, that is 0.64. The refractive indices $n_{\text{Si}} = 1.45$ and $n_{\text{air}} = 1$ have been assumed for the silica and air refractive index, respectively, and the chromatic dispersion of the silica has been taken into account by calculating the refractive index through the Sellmeier equation [2.42].

Figure 2.30 shows the dispersion curves of the fundamental and the higher-order modes of the four fibers here considered. All the dispersion curves, which are inside the bandgap for a wide wavelength range, are always under the air-line, defined by $n_{\text{eff}} = 1$, as required for the air-guiding. Notice that all

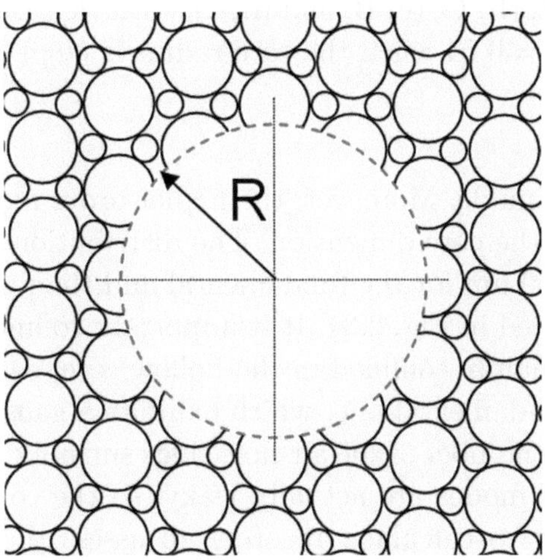


Figure **2.29**: PBGF cross-section with the hollow-core radius $R$. $R = 2\Lambda$ for fiber A and C, while $R = 3\Lambda$ for fiber B and D [2.7].

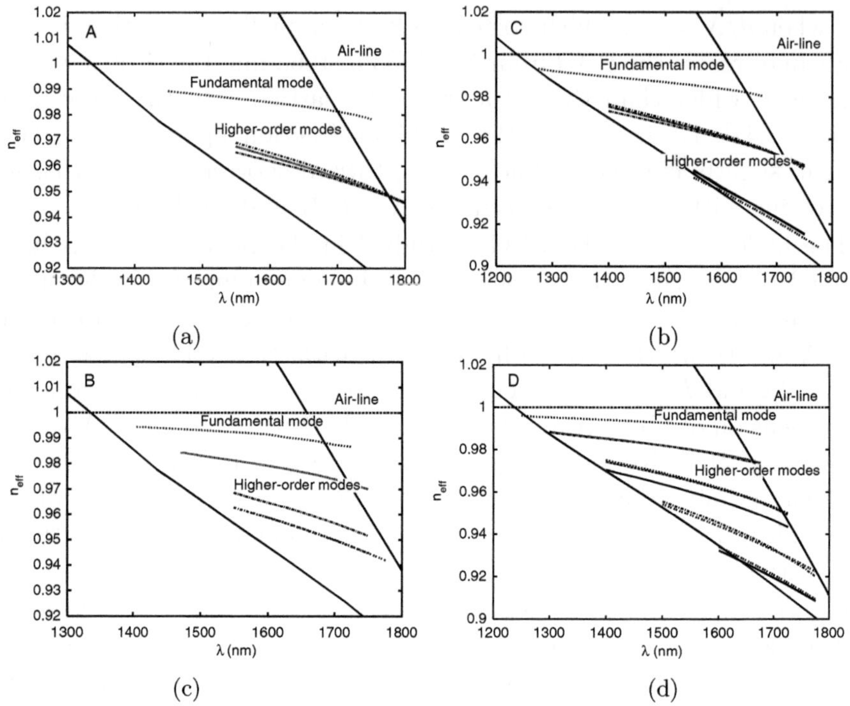


Figure **2.30**: Dispersion curves of the fundamental and the higher-order modes of the PBGFs (a) A, (b) C, (c) B, and (d) D, with (*left column*) $d/\Lambda = 0.6$ and (*right column*) $d/\Lambda = 0.64$ when the core radius is (*top*) $R = 2\Lambda$ and (*bottom*) $R = 3\Lambda$ [2.7].

the fibers are multi-mode. Moreover, the higher-order mode number increases with $d/\Lambda$ and with the core dimension. The distribution of the magnetic field modulus at $\lambda = 1550$ nm for the fundamental and the first higher-order mode of fiber A are reported in Fig. 2.31. It is important to underline that both the guided modes are mainly confined in the hollow core, even if the confinement is tighter for the fundamental one, which exhibits a gaussian-like shape.

Due to the finite number of the air-hole rings surrounding the PBGF hollow core, all the guided modes are actually leaky, so the confinement loss $CL$ is a crucial parameter to calculate, in order to assess the applicability of the fibers with modified honeycomb lattice here presented. This loss, related to the leakage phenomenon, are defined, as usual, according to Eq. (A.9). As it has been already demonstrated, $CL$ strongly depends on the ring number and

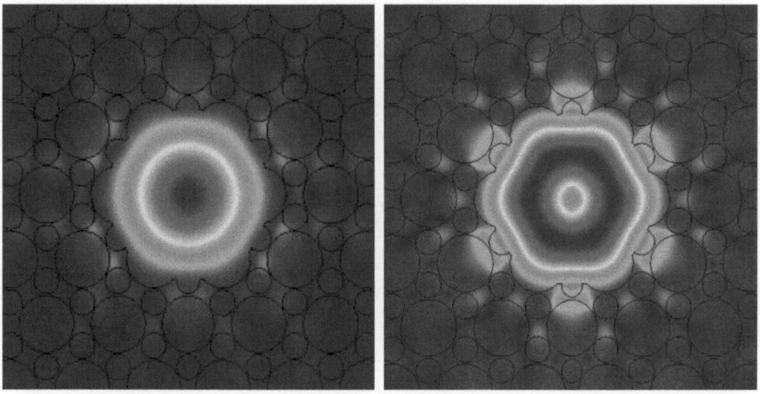

(a) (b)

Figure **2.31**: Magnetic field modulus of (a) the fundamental and (b) the first higher-order mode of fiber A at $\lambda = 1550$ nm [2.7].

on the wavelength [2.43]. In the present study PBGFs with eight unitary cells around the hollow core have been considered. Figure 2.32 shows the spectral behavior of the $CL$ for the fundamental and the higher-order modes of the four fibers. It is possible to notice that all the $CL$ curves exhibit the U-shape typical of PBGFs. As expected, the higher-order modes present higher $CL$ with respect to the fundamental one, due to the lower field confinement. It is important to underline that the $CL$ minimum for the higher-order modes falls at longer wavelengths with respect to the fundamental mode one. In fact, as it can be observed from Fig. 2.30, the higher-order modes present a lower effective index than the fundamental one, thus the crossing of the PBG center is at a longer wavelength. For the same reason, the $CL$ minima shift towards shorter wavelengths when $d/\Lambda$ increases from 0.6 to 0.64. Focusing on the fundamental mode $CL$, the minimum becomes lower by increasing both the core radius $R$ and the air-hole diameter $d$, that is the $d/\Lambda$ value for a fixed pitch. The reduction of the confinement loss due to the larger core radius is higher when $d/\Lambda = 0.64$. For example, with $d/\Lambda = 0.6$ the $CL$ minimum changes from 12 dB/km at 1575 nm for fiber A to 0.9 dB/km at 1550 nm for fiber B, with a decrease of one order of magnitude. When $d/\Lambda = 0.64$, the difference between the $CL$ minimum for fibers C and D is more than two orders of magnitude, being $2 \cdot 10^{-1}$ dB/km at 1450 nm for fiber C and $5.5 \cdot 10^{-3}$ dB/km at 1425 nm for fiber D. Notice that only the $CL$ values of the last PBGF are lower than the Rayleigh scattering limit. However,

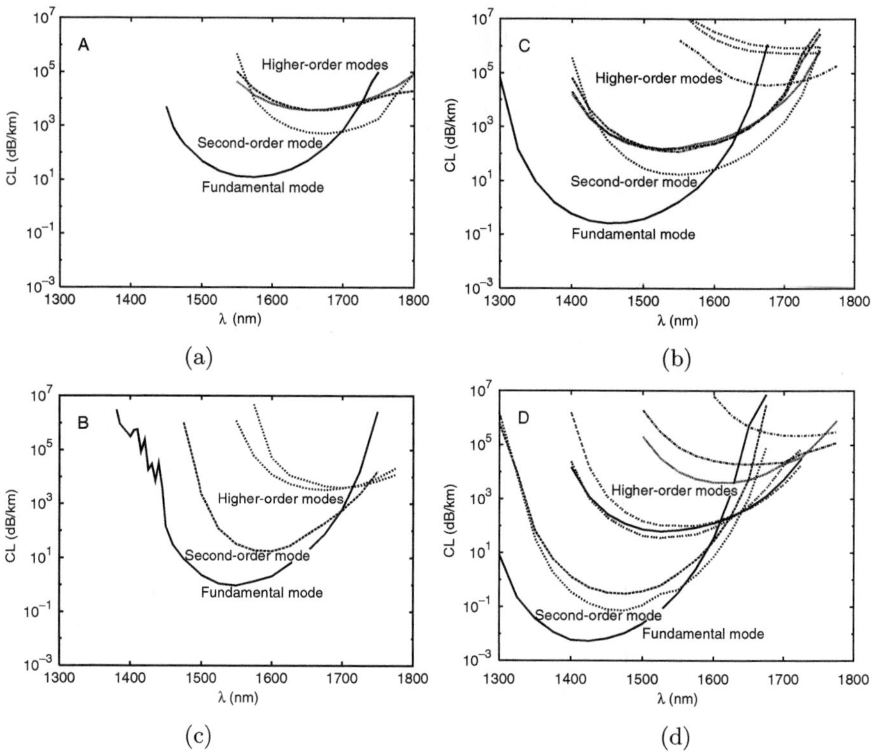


Figure **2.32**: Confinement loss versus the wavelength of the fundamental and the higher-order modes of the PBGFs (a) A, (b) C, (c) B, and (d) D, with (*left column*) $d/\Lambda = 0.6$ and (*right column*) $d/\Lambda = 0.64$ when the core radius is (*top*) $R = 2\Lambda$ and (*bottom*) $R = 3\Lambda$ [2.7].

by increasing the number of the air-hole rings surrounding the hollow core, it is possible to further decrease the $CL$ values of all the proposed fibers, making them negligible not only at the PBG center, but in a wider wavelength range [2.43, 2.6].

### 2.3.2 Birefringence

As already stated, another interesting property of hollow-core PCFs is related to the phase-index and group-index birefringence, which have been investigated both experimentally [2.44] and numerically [2.45–2.47], as already done for solid-core PCFs [2.48–2.50]. The high refractive index contrast and the great flexibility of the fabrication process of microstructured fibers allow to obtain a birefringence of at least one order of magnitude higher than that of standard birefringent ones, such as PANDA and bow tie fibers, which usually show a modal birefringence of the order of $10^{-4}$ [2.46]. These high birefringence values have been usually obtained with a proper asymmetric central defect design in both solid and hollow-core PCFs, or by changing the size and the shape of the air-holes surrounding the fiber silica core along the two orthogonal axes [2.44, 2.46].

All the hollow-core PBGFs with high birefringence proposed in literature [2.45, 2.47] and realized [2.44] are characterized by an unitary cell with a high air-filling fraction, which results from the presence of big air-holes separated by very thin silica bridges, often seat of undesired surface modes. In addition, it has been demonstrated that asymmetries in the silica ring surrounding the hollow core of PBGFs, where these surface modes are located, can strongly influence the fiber polarization properties [2.51]. Moreover, the photonic crystal cladding extension determines the amount of the confinement loss of the field of the guided modes and, in turn, whether or not they can actually propagate. Thus, in order to obtain a birefringent PBGF to be used in real applications, it is mandatory to analyze the losses and the field distribution of the modes sustained by the fiber, that is to identify proper lattice structures and hollow-core designs which allow to avoid surface modes, to minimize the fundamental mode confinement loss and to obtain an effectively single-mode operation region.

To this aim, new hollow-core geometries have been considered in the modified honeycomb PBGFs previously described, providing birefringence values up to $10^{-3}$, as well as the absence of surface modes [2.52], an effectively single-mode behavior in C and L bands, and fundamental mode confinement loss lower than 1 dB/km in the same wide wavelength range. All the highly birefringent PBGFs here designed are characterized by $d/\Lambda = 0.6$, $d_c/\Lambda = 1.32$ and $\Lambda = 1.62$ μm. As shown in Fig. 2.28 in Section 2.3, the PBG obtained with these geometric parameter values is crossed by the air-line from $\lambda = 1333$ to 1663 nm [2.43].

The first considered hollow core, reported in Fig. 2.33, has been obtained by removing eight lattice unitary cells around the central one in the fiber cross-section, thus introducing a geometric asymmetry. The modulus of the fundamental mode magnetic field main component at 1550 nm, that is at the normalized wavelength $\lambda/\Lambda \simeq 0.957$, is shown in Fig. 2.34 for the two

Figure **2.33**: Cross-section of a highly birefringent PBGF.

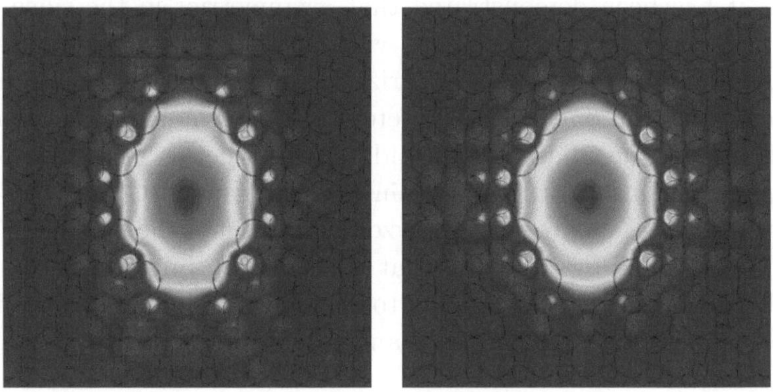

(a) (b)

Figure **2.34**: Magnetic field modulus of the fundamental mode main component, (a) $x$-polarized and (b) $y$-polarized, at $\lambda = 1550$ nm.

polarizations. It is possible to notice that the fundamental mode is strongly confined in air, even if there is still some field in the silica regions surrounding the hollow core.

The dispersion curve for the $x$ and $y$ polarization of the fundamental mode, which is guided inside the PBG approximately from 1425 to 1750 nm, is reported in Fig. 2.35a. The phase-index birefringence $B = |n_{\text{eff},x} - n_{\text{eff},y}|$ has been calculated starting from the effective index values and it is shown as a function of the normalized wavelength $\lambda/\Lambda$ in Fig. 2.35b. It is important to underline that the birefringence $B$ is higher than $1 \cdot 10^{-4}$ in a wavelength

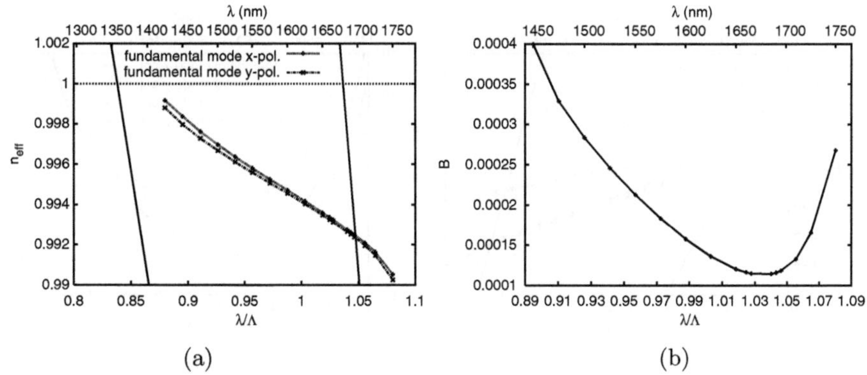


Figure **2.35**: (a) Dispersion curve of the two polarizations of the fundamental mode and (b) birefringence as a function of the normalized wavelength.

range of about 300 nm, reaching a maximum of $4 \cdot 10^{-4}$ at $\lambda/\Lambda \simeq 0.895$, that is at 1450 nm, and then decreasing towards a minimum value of $1.14 \cdot 10^{-4}$ at $\lambda/\Lambda \simeq 1.04$, that is at 1685 nm. Finally, $B$ increases again to $2.675 \cdot 10^{-4}$ at the longest wavelength of the considered range. Notice that the birefringence obtained with this modified honeycomb PBGF, which is $2.13 \cdot 10^{-4}$ at 1550 nm, is similar to that of the conventional polarization maintaining fibers [2.46]. However, the hollow-core fiber proposed presents all the advantages related to the light propagation in air.

The studied hollow-core birefringent fiber is surface-mode free. However, it is multi-mode, since higher-order modes coexist with the fundamental one in the PBG. Looking at the first higher-order mode dispersion curve in Fig. 2.36a, it is possible to notice that the coupling with the fundamental mode is weak, being the effective index difference between $2 \cdot 10^{-3}$ and $8.6 \cdot 10^{-3}$ in the $\lambda/\Lambda$ range between 0.895 and 1.08. In order to evaluate the first higher-order mode influence on the propagation of the fundamental one, it is useful to calculate the confinement loss, according to Eq. (A.9) [2.43]. In fact, if the second-order mode $CL$ values are high enough, only the fundamental mode can propagate along the fiber. In particular, Fig. 2.36b reports the $CL$ values as a function of the normalized wavelengths for these two guided modes in the designed PCF, which is characterized by eleven air-hole rings. As expected, the field confinement in the hollow core is weaker for the first higher-order mode. In fact, its $CL$ values are at least one order of magnitude higher than those of the fundamental one for both the polarizations in all the considered wavelength

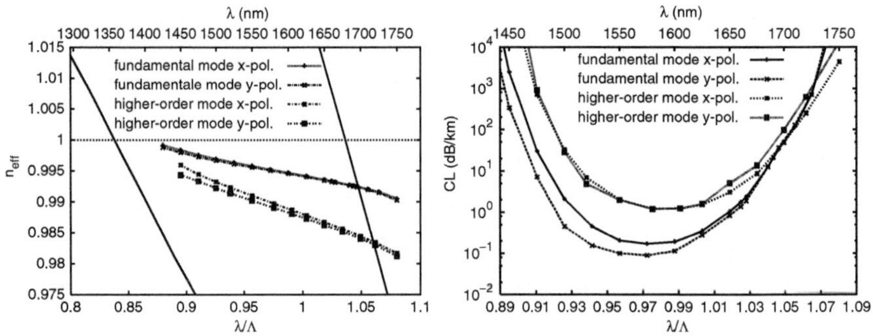


(a) (b)

Figure **2.36**: (a) Dispersion curve and (b) confinement loss for the two polarizations of the fundamental and the higher-order mode as a function of the normalized wavelength.

range. In particular, the $CL$ minimum is about 1.23 dB/km at 1605 nm and 1.2 dB/km at 1580 nm for the $x$ and $y$ polarization, respectively. As regards the fundamental mode, its losses present the typical U-shape behavior with a minimum of 0.17 and 0.09 dB/km at 1575 nm for $x$ and $y$ polarization, respectively.

Starting from the previous considerations on the coupling efficiency and the confinement loss of the first higher-order mode, the proposed birefringent PBGF can be considered effectively single-mode in the wavelength range of interest. Moreover, it is important to underline that the fundamental mode $CL$ are negligible, that is lower than the limit of 0.2 dB/km fixed by the Rayleigh scattering, in a 50 nm wavelength range, between 1550 nm and 1600 nm, for the $x$ polarization, and in a 100 nm one, centered around 1565 nm, for the $y$ polarization. In particular, the designed fiber presents a birefringence value between $1.58 \cdot 10^{-4}$ and $2.13 \cdot 10^{-4}$ in the 50 nm wavelength range where it is effectively single mode.

A second kind of hollow-core asymmetry has been introduced in the modified honeycomb PBGFs by removing the silica inside an ellipse with minor semiaxis $a$ and major semiaxis $b$ along the $x$ and $y$ direction, respectively. Results regarding the three fibers shown in Fig. 2.37, called fibers A, B, and C, are here discussed. In particular, for PBGF A $a = \sqrt{3}\Lambda \simeq 2.806$ μm and $b = 3\Lambda \simeq 4.86$ μm, while the hollow cores of fibers B and C are slightly smaller, being $a = 2.5$ μm and $b = 4.5$ μm, and $a = 2.4$ μm and $b = 4.4$ μm,

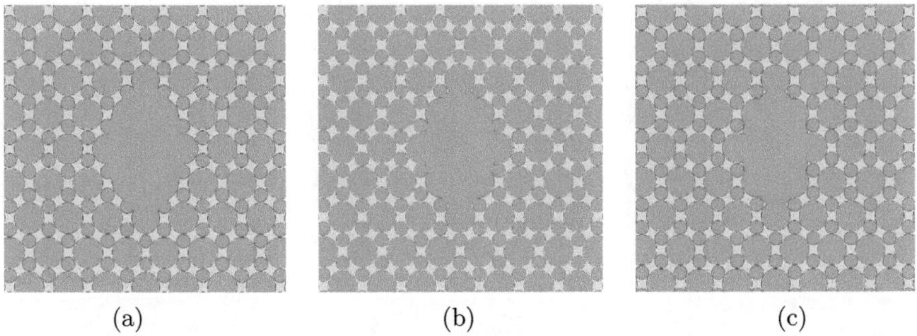

Figure **2.37**: Cross-section of the highly birefringent PBGF (a) A, (b) B, and (c) C.

respectively. Notice that, differently from the other two fibers, eight air-holes of diameter $d$ are completely excluded from the hollow core of the PBGF C, which is the smallest one.

Notice that the distribution of the magnetic field modulus at $\lambda = 1550$ nm for the two polarizations of the fundamental mode, reported in Fig. 2.38 for all the considered PBGFs, is mainly confined in the hollow core. Differences in the guided-mode field distribution can be noticed, which are due to the different hollow-core geometry of the three fibers.

The dispersion curves of the two polarizations of the fundamental and the first higher-order mode, as well as the PBG edges, are reported in Fig. 2.39 for the three designed PBGFs. It is important to underline the absence of surface modes, which, on the contrary, strongly affect the realistic highly birefringent PBGF previously studied [2.46], thus making difficult to distinguish the fundamental mode at some wavelengths. Notice that avoiding the surface-mode presence prevents also the influence of the small structural features and distorsions of the silica ring surrounding the fiber hollow core on the polarization behavior of the PBGFs [2.51].

As already stated, a single-mode operation regime for the PBGFs is desirable, since the higher-order modes, if excited, can negatively affect the fiber polarization properties [2.44]. Unfortunately, as reported in Fig. 2.39, all the modified honeycomb birefringent PBGFs here designed are multi-mode. However, notice that the coupling between the fundamental and the higher-order mode is very weak for the proposed PBGFs, being the effective index difference higher than 0.01, 0.009, and 0.006 for fiber A, B, and C, respectively, in the considered wavelength range.

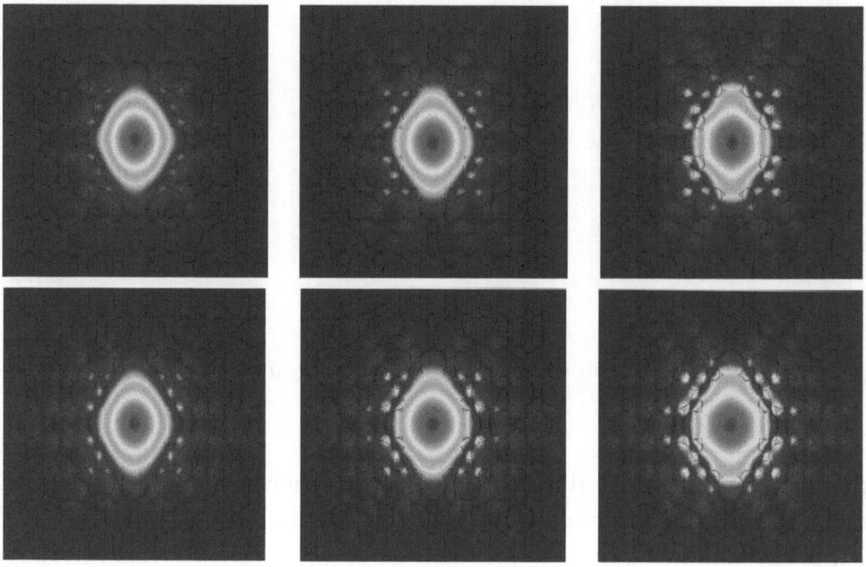

(a) (b) (c)

Figure **2.38**: Magnetic field modulus of (*top*) the $x$-polarization and (*bottom*) the $y$-polarization of the fundamental mode at $\lambda = 1550$ nm for the highly birefringent PBGF (a) A, (b) B, and (c) C.

The single-mode regime of the modified honeycomb fibers here proposed have been again investigated by calculating the confinement loss of the fundamental and of the higher-order modes. In particular, a PBGF is considered to be effectively single mode in the wavelength range where the fundamental mode confinement loss is one order of magnitude lower than the $CL$ minimum of the first higher-order one. Figure 2.40 shows the spectral behavior of the $CL$ for the fundamental and the higher-order modes for fibers A, B, and C. Notice that, as expected, the second-order mode $CL$ is higher than that of the fundamental one for all the three highly birefringent fibers. Moreover, it is important to underline that the $CL$ minimum for the higher-order modes falls at longer wavelengths with respect to the fundamental mode one, which is about 0.17 and 0.22 dB/km at 1600 nm for both the polarizations of PBGF A and B, respectively. For fiber C there is a slightly higher difference between the $CL$ minimum values for the two polarizations, being 0.28 dB/km at 1600 nm and 0.19 dB/km at 1575 nm. According to the previous definition, it has been demonstrated that fiber A can be considered effectively single-mode from

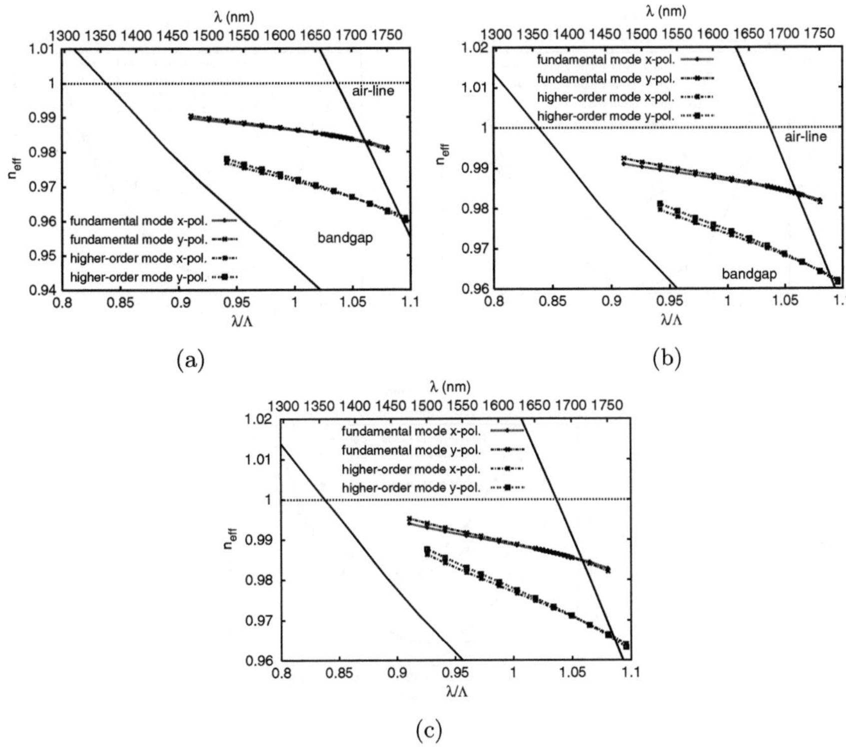


Figure **2.39**: Dispersion curves for the two polarizations of fundamental and first higher-order modes of the highly birefringent PBGF (a) A, (b) B, and (c) C.

1525 to 1660 nm, fiber B between 1535 and 1665 nm, and fiber C from 1550 to 1645 nm, that is in a wavelength range of 135, 130, and 95 nm, respectively. Moreover, notice that in these wavelength ranges all the PBGFs with elliptical hollow core present fundamental mode $CL$ almost lower than 1 dB/km, which becomes almost negligible around the PBG center.

Starting from the previous analysis, it is useful to consider the phase birefringence $B$, shown in Fig. 2.41 for the three designed PBGFs, only in the wavelength range where the fibers are effectively single mode. The birefringence value decreases for all the hollow-core fibers as the wavelength increases, starting from a maximum value of about $5.8 \cdot 10^{-4}$ 1525 nm, $1 \cdot 10^{-3}$ at 1535 nm and $7.2 \cdot 10^{-4}$ at 1550 nm for PBGF A, B, and C, respectively. Notice that the $B$ values are higher for the fiber with the hollow core of intermediate

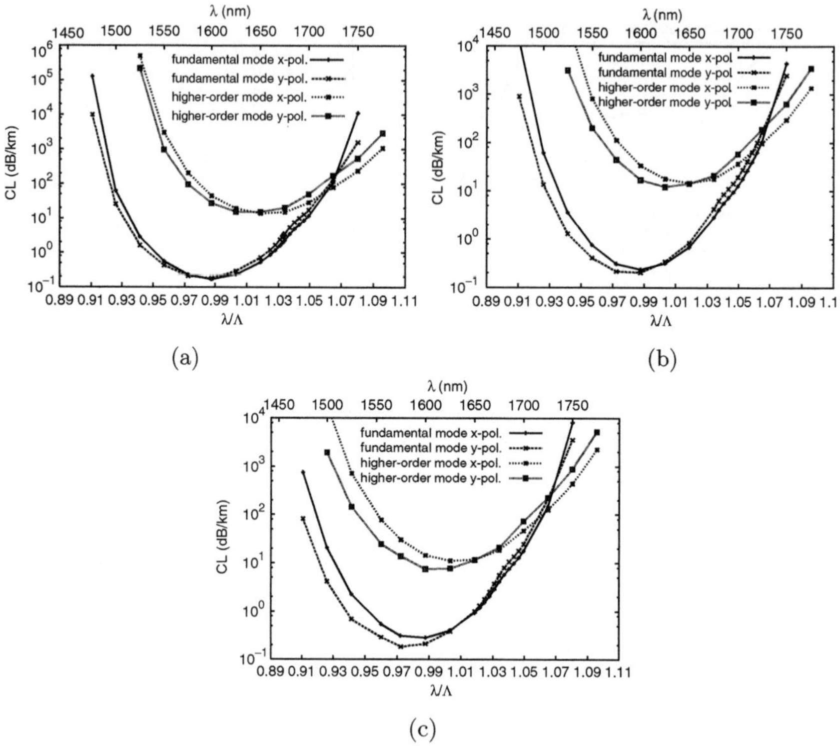


Figure **2.40**: Confinement loss for the two polarizations of fundamental and first higher-order modes of the highly birefringent PBGF (a) A, (b) B, and (c) C.

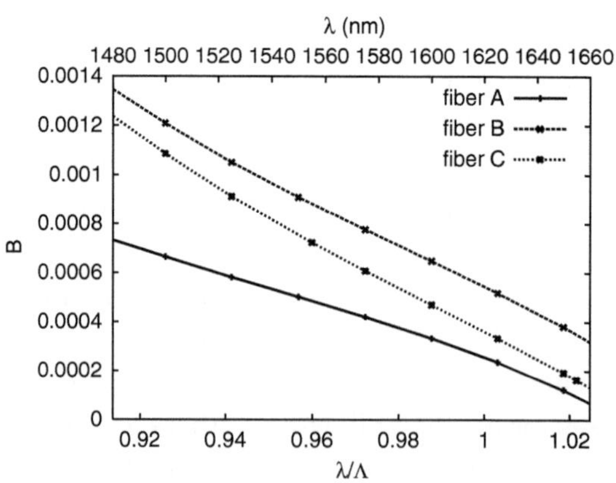


Figure **2.41**: Birefringence as a function of the wavelength for the highly birefringent PBGFs A, B, and C.

dimension, that is for fiber B, being almost $3 \cdot 10^{-4}$ in a wavelength range of about 120 nm, that is all over the C and L bands. A further decrease of the hollow-core dimension, as in fiber C, does not provide better birefringence properties. In fact, simulation results have shown that intermediate birefringence values with respect to fibers A and B can be obtained with PBGF C, as well as a smaller effectively single-mode range.

# Bibliography


[2.1] A. H. Bouk, A. Cucinotta, F. Poli, and S. Selleri, "Dispersion properties of square-lattice photonic crystal fibers," *Optics Express*, vol. 12, pp. 941–946, Mar. 2004. Available at: http://www.opticsexpress.org/abstract.cfm?URI=OPEX-12-5-941

[2.2] F. Poli, M. Foroni, M. Bottacini, M. Fuochi, N. Burani, L. Rosa, A. Cucinotta, and S. Selleri, "Single-mode regime of square-lattice photonic crystal fibers," *Journal of Optical Society of America A*, vol. 22, pp. 1655–1661, Aug. 2005.

[2.3] M. Foroni, F. Poli, L. Rosa, A. Cucinotta, and S. Selleri, "Cut-off properties of large-mode-area photonic crystal fibers," in *Proc. IEEE/LEOS Workshop on Fibres and Optical Passive Components WFOPC 2005*, Palermo, Italy, June 22–24, 2005.

[2.4] S. Selleri, A. Cucinotta, M. Foroni, F. Poli, and M. Bottacini, "New design of single-mode large-mode-area photonic crystal fibers," in *Proc. International Congress on Optics and Optoelectronics SPIE-COO 2005*, Warsaw, Poland, Aug. 28–Sept. 2, 2005.

[2.5] L. Vincetti, F. Poli, A. Cucinotta, and S. Selleri, "Wide bandgap air-guiding modified honeycomb photonic crystal fibers," in *Proc. CLEO Europe 2005*, Munich, Germany, June 12–17, 2005.

[2.6] S. Selleri, L. Vincetti, F. Poli, A. Cucinotta, and M. Foroni, "Air-guiding photonic crystal fibers with modified honeycomb lattice," in *Proc. IEEE/LEOS Workshop on Fibres and Optical Passive Components WFOPC 2005*, Palermo, Italy, June 22–24, 2005.

[2.7] L. Vincetti, M. Maini, F. Poli, A. Cucinotta, and S. Selleri, "Numerical analysis of hollow core photonic band gap fibers with modified honeycomb lattice," *Optical and Quantum Electronics*, Dec. 2006.

[2.8] P. St. J. Russell, E. Marin, A. Díez, S. Guenneau, and A. B. Movchan, "Sonic band gaps in PCF preforms: enhancing the interaction of sound and light," *Optics Express*, vol. 11, pp. 2555–2560, Oct. 2003. Available at: http://www.opticsexpress.org/abstract.cfm?URI=OPEX-11-20-2555

[2.9] M. G. Franczyk, J. C. Knight, T. A. Birks, P. St. J. Russell, and A. Ferrando, "Birefringent photonic crystal fiber with square lattice," in *Lightguides and their Applications II*, J. Wojcik and W. Wojcik, Eds. Proc. SPIE, 2004, vol. 5576, pp. 25–28.

[2.10] Y. C. Liu and Y. Lai, "Optical birefringence and polarization dependent loss of square- and rectangular-lattice holey fibers with elliptical air holes: numerical analysis," *Optics Express*, vol. 13, pp. 225–235, Jan. 2005. Available at: http://www.opticsexpress.org/abstract.cfm?URI=OPEX-13-1-225

[2.11] F. Poli, A. Cucinotta, M. Fuochi, S. Selleri, and L. Vincetti, "Characterization of microstructured optical fibers for wideband dispersion compensation," *Journal of Optical Society of America A*, vol. 20, pp. 1958–1962, Oct. 2003.

[2.12] A. Cucinotta, S. Selleri, L. Vincetti, and M. Zoboli, "Holey fiber analysis through the finite-element method," *IEEE Photonics Technology Letters*, vol. 14, pp. 1530–1532, Nov. 2002.

[2.13] A. Cucinotta, F. Poli, S. Selleri, L. Vincetti, and M. Zoboli, "Amplification properties of $Er^{3+}$-doped photonic crystal fibers," *IEEE/OSA Journal of Lightwave Technology*, vol. 21, pp. 782–788, Mar. 2003.

[2.14] B. T. Kuhlmey, R. C. McPhedran, C. M. de Sterke, P. A. Robinson, G. Renversez, and D. Maystre, "Microstructured optical fibers: where's the edge?," *Optics Express*, vol. 10, pp. 1285–1290, Nov. 2002. Available at: http://www.opticsexpress.org/abstract.cfm?URI=OPEX-10-22-1285

[2.15] N. A. Mortensen, J. R. Folkenberg, M. D. Nielsen, and K. P. Hansen, "Modal cutoff and the V parameter in photonic crystal fibers," *Optics Letters*, vol. 28, pp. 1879–1881, Oct. 2003.

[2.16] J. C. Knight, T. A. Birks, P. St. J. Russell, and D. M. Atkin, "All-silica single-mode optical fiber with photonic crystal cladding," *Optics Letters*, vol. 21, pp. 1547–1549, Oct. 1996.

[2.17] T. A. Birks, J. C. Knight, and P. St. J. Russell, "Endlessly single-mode photonic crystal fiber," *Optics Letters*, vol. 22, pp. 961–963, July 1997.

[2.18] D. Ferrarini, L. Vincetti, M. Zoboli, A. Cucinotta, and S. Selleri, "Leakage properties of photonic crystal fibers," *Optics Express*, vol. 10, pp. 1314–1319, Nov. 2002. Available at: http://www.opticsexpress.org/abstract.cfm?URI=OPEX-10-23-1314

[2.19] B. Kuhlmey, G. Renversez, and D. Maystre, "Chromatic dispersion and losses of microstructured optical fibers," *Applied Optics*, vol. 42, pp. 634–639, Feb. 2003.

[2.20] L. Vincetti, "Confinement losses in honeycomb fibers," *IEEE Photonics Technology Letters*, vol. 16, pp. 2048–2050, Sept. 2004.

[2.21] B. T. Kuhlmey, R. C. McPhedran, and C. M. de Sterke, "Modal cutoff in microstructured optical fibers," *Optics Letters*, vol. 27, pp. 1684–1686, Oct. 2002.

[2.22] N. A. Mortensen, "Effective area of photonic crystal fiber," *Optics Express*, vol. 10, pp. 341–348, Apr. 2002. Available at: http://www.opticsexpress.org/abstract.cfm?URI=OPEX-10-7-341

[2.23] T. P. White, B. T. Kuhlmey, R. C. McPhedran, D. Maystre, G. Renversez, C. M. de Sterke, and L. Botten, "Multipole method for microstructured optical fibers I. Formulation," *Journal of Optical Society of America B*, vol. 19, pp. 2322–2330, Oct. 2002.

[2.24] T. P. White, B. T. Kuhlmey, R. C. McPhedran, D. Maystre, G. Renversez, C. M. de Sterke, and L. Botten, "Multipole method for microstructured optical fibers II. Implementation and results," *Journal of Optical Society of America B*, vol. 19, pp. 2331–2340, Oct. 2002.

[2.25] T. A. Birks, D. Mogilevstev, J. C. Knight, P. St. J. Russell, J. Broeng, P. J. Roberts, J. A. West, D. J. Allan, and J. C. Fajardo, "The analogy between photonic crystal fibres and step index fibres," in *Proc. Optical Fiber Communications Conference OFC 1999*, Feb. 21–26, 1999, paper FG4-1.

[2.26] F. Brechet, J. Marcou, D. Pagnoux, and P. Roy, "Complete analysis of the characteristics of propagation into photonic crystal fibers, by the finite element method," *Optical Fiber Technology*, vol. 6, pp. 181–191, Apr. 2000.

[2.27] M. Koshiba, "Full-vector analysis of photonic crystal fibers using the finite element method," *IEICE Transactions on Electronics*, vol. E85-C, pp. 881–888, Apr. 2002.

[2.28] M. Koshiba and K. Saitoh, "Applicability of classical optical fiber theories to holey fibers," *Optics Letters*, vol. 29, pp. 1739–1741, Aug. 2004.

[2.29] M. D. Nielsen and N. A. Mortensen, "Photonic crystal fiber design based on the V-parameter," *Optics Express*, vol. 11, pp. 2762–2767, Oct. 2003. Available at: http://www.opticsexpress.org/abstract.cfm?URI=OPEX-11-21-2762

[2.30] S. G. Johnson and J. D. Joannopoulos, "Block-iterative frequency-domain methods for Maxwell's equations in a planewave basis," *Optics Express*, vol. 8, pp. 173–179, Jan. 2001. Available at: http://www.opticsexpress.org/abstract.cfm?URI=OPEX-8-3-173

[2.31] A. Ferrando, E. Silvestre, J. J. Miret, and P. Andrés, "Full-vector analysis of a realistic photonic crystal fiber," *Journal of Optical Society of America A*, vol. 17, pp. 1333–1340, July 2000.

[2.32] J. C. Baggett, T. M. Monro, K. Furusawa, and D. J. Richardson, "Comparative study of large-mode holey and conventional fibers," *Optics Letters*, vol. 26, pp. 1045–1047, July 2001.

[2.33] N. A. Mortensen, M. D. Nielsen, J. R. Folkenberg, A. Petersson, and H. Simonsen, "Improved large-mode-area endlessly single-mode photonic crystal fibers," *Optics Letters*, vol. 28, pp. 393–395, Mar. 2003.

[2.34] M. Nielsen, N. A. Mortensen, J. Folkenberg, A. Petersson, and A. Bjarklev, "Improved all-silica endlessly single-mode photonic crystal fiber," in *Proc. Optical Fiber Communications Conference OFC 2003*, Atlanta, Georgia, USA, Mar. 23–28, 2003.

[2.35] C. M. Smith, N. Venkataraman, M. T. Gallagher, D. Müller, J. A. West, N. F. Borrelli, D. C. Allan, and K. W. Koch, "Low-loss hollow-core silica/air photonic bandgap fibre," *Nature*, vol. 424, pp. 657–659, Aug. 2003.

[2.36] Y. Xu and A. Yariv, "Loss analysis of air-core photonic crystal fibers," *Optics Letters*, vol. 28, pp. 1885–1887, Oct. 2003.

[2.37] M. Yan and P. Shum, "Air guiding with honeycomb photonic bandgap fiber," *IEEE Photonics Technology Letters*, vol. 17, pp. 64–66, Jan. 2005.

[2.38] M. Yan, P. Shum, and J. Hu, "Design of air-guiding honeycomb photonic bandgap fiber," *Optics Letters*, vol. 30, pp. 465–467, Mar. 2005.

[2.39] T. Murao, K. Saitoh, and M. Koshiba, "Design of air-guiding modified honeycomb photonic band-gap fibers for effectively singlemode operation," *Optics Express*, vol. 14, pp. 2404–2412, Mar. 2006. Available at: http://www.opticsexpress.org/abstract.cfm?URI=oe-14-6-2404

[2.40] M. Chen and R. Yu, "Analysis of photonic bandgaps in modified honeycomb structures," *IEEE Photonics Technology Letters*, vol. 16, pp. 819–821, Jan. 2004.

[2.41] K. Saitoh and M. Koshiba, "Confinement losses in air-guiding photonic bandgap fibers," *IEEE Photonics Technology Letters*, vol. 15, pp. 236–238, Feb. 2003.

[2.42] G. P. Agrawal, *Nonlinear Fiber Optics*. New York: Academic, 2001.

[2.43] L. Vincetti, F. Poli, and S. Selleri, "Confinement loss and nonlinearity analysis of air-guiding modified honeycomb photonic bandgap fibers," *IEEE Photonics Technology Letters*, vol. 18, pp. 508–510, Feb. 2006.

[2.44] X. Chen, M. Li, N. Venkataraman, M. Gallagher, W. Wood, A. Crowley, J. Carberry, L. Zenteno, and K. W. Koch, "Highly birefringent hollow-core photonic bandgap fiber," *Optics Express*, vol. 12, pp. 3888–3893, Aug. 2004. Available at: http://www.opticsexpress.org/abstract.cfm?URI=oe-12-16-3888

[2.45] K. Saitoh and M. Koshiba, "Photonic bandgap fibers with high birefringence," *IEEE Photonics Technology Letters*, vol. 14, pp. 1291–1293, Sept. 2002.

[2.46] M. S. Alam, K. Saitoh, and M. Koshiba, "High group birefringence in air-core photonic bandgap fibers," *Optics Letters*, vol. 30, pp. 824–826, Apr. 2005.

[2.47] M. Szpulak, R. Kotynski, T. Nasilowski, W. Urbańczyk, and H. Thienpont, "Form birefringence of air guiding photonic crystal fibers," in *Proc. 9th Ann. Symposium IEEE/LEOS Benelux Chapter*, Los Angeles, California, USA, Dec. 2004, pp. 319–322.

[2.48] A. Ortigosa-Blanch, J. C. Knight, W. J. Wadsworth, J. Arriaga, B. J. Mangan, T. A. Birks, and P. St. J. Russell, "Highly birefringent photonic crystal fibers," *Optics Letters*, vol. 25, pp. 1325–1327, Sept. 2000.

[2.49] T. P. Hansen, J. Broeng, S. E. B. Libori, E. Knudsen, A. Bjarklev, J. R. Jensen, and H. Simonsen, "Highly birefringent index-guiding photonic crystal fibers," *IEEE Photonics Technology Letters*, vol. 13, pp. 588–590, June 2001.

[2.50] C. L. Zhao, X. F. Yang, C. Lu, W. Jin, and M. S. Demokan, "Temperature-insensitive interferometer using a highly birefringent photonic crystal fiber loop mirror," *IEEE Photonics Technology Letters*, vol. 16, pp. 2535–2537, Nov. 2004.

[2.51] F. Poletti, N. G. R. Broderick, D. J. Richardson, and T. M. Monro, "The effect of core asymmetries on the polarization properties of hollow core photonic bandgap fibers," *Optics Express*, vol. 13, pp. 9115–9124, Oct. 2005. Available at: http://www.opticsexpress.org/abstract.cfm?URI=oe-13-22-9115

[2.52] J. A. West, C. M. Smith, N. F. Borrelli, D. C. Allan, and K. W. Koch, "Surface modes in air-core photonic band-gap fibers," *Optics Express*, vol. 12, pp. 1485–1496, Apr. 2004. Available at: http://www.opticsexpress.org/abstract.cfm?URI=oe-12-8-1485

# Chapter 3

# Dispersion properties

In this chapter results regarding the PCF dispersion properties are reported. The analyses performed have shown that, by properly changing the geometric characteristics of the air-holes in the PCF cross-section, the waveguide contribution to the dispersion parameter can be significantly changed, thus obtaining unusual positions of the zero-dispersion wavelength, as well as particular values of the dispersion curve slope. In particular, by manipulating the air-hole radius or the lattice period of the microstructured cladding, it is possible to control the zero-dispersion wavelength, which can be tuned over a very wide range [3.1–3.3], or the dispersion curves, which can be engineered to be ultraflattened [3.4–3.7].

First of all, it is reported the study of the dispersion properties of triangular PCFs with a high air-filling fraction, that is with small hole-to-hole spacing and large air-holes, which can be designed to compensate the anomalous dispersion and the dispersion slope of single-mode fibers [3.8–3.10]. In particular, the geometric parameters which characterize these triangular PCFs have been chosen to optimize the fiber length and the dispersion compensation over a wide wavelength range.

Then the dispersion properties of PCFs with a square lattice of air-holes have been investigated for different values of the geometric parameters which describe the fiber cross-section. In particular, large air-holes and small pitch have been considered, in order to make a comparison with the dispersion curves of the triangular PCFs with the same $\Lambda$ and $d/\Lambda$ values [3.11]. It has been demonstrated that also with this air-hole disposition, that is with the square lattice, negative values of the dispersion parameter and of the dispersion slope can been obtained in the wavelength range centered at 1550 nm.

In the second part of the chapter the design of triangular PCFs with completely different characteristics, that is with flattened dispersion curve and zero-dispersion wavelength around 1550 nm, which can be exploited for nonlinear applications, is described. The triangular PCF cross-section geometry around the core has been modified in two different ways, in order to obtain the desired dispersion properties and a small effective area, that is a high nonlinear coefficient. In the first fiber type the diameter of the air-holes belonging to the first three rings has been properly changed, that is their dimension has been decreased or increased [3.12]. On the contrary, in the second PCF type the central air-hole has been removed and the diameter of three air-holes belonging to the first ring has been reduced, thus obtaining a silica core with a triangular shape [3.13, 3.14]. Results have demonstrated that it is possible to successfully design highly nonlinear triangular PCF with effective area of few $\mu m^2$, flattened dispersion curve, and zero-dispersion wavelength in the C band with both the core configurations here considered.

## 3.1 PCFs for dispersion compensation

PCFs with a high air-filling fraction have been designed in order to compensate the anomalous dispersion and the dispersion slope of SMFs. In fact, their chromatic dispersion limits the data transmission rate in broadband wavelength division multiplexing (WDM) systems. In particular, it becomes a critical issue as soon as the transmission bit-rate increases over 10 Gb/s.

The positive dispersion of installed fibers can be compensated by dispersion compensating fibers (DCFs) with a large dispersion of opposite sign. For WDM systems this goal must be achieved over a broad wavelength range around 1550 nm, thus implying, besides large negative dispersion values, a proper negative dispersion slope.

The present analysis has demonstrated that PCFs can be exploited to this aim. In fact, their dispersion properties can be modified with high flexibility, since the large refractive index variation between silica and air permits to achieve a significant waveguide dispersion over a wide wavelength range. PCFs with large air-holes have been already proposed in literature for dispersion compensation, even though their description has been performed through a simplified model consisting of a silica core in air [3.15]. When the wavelength

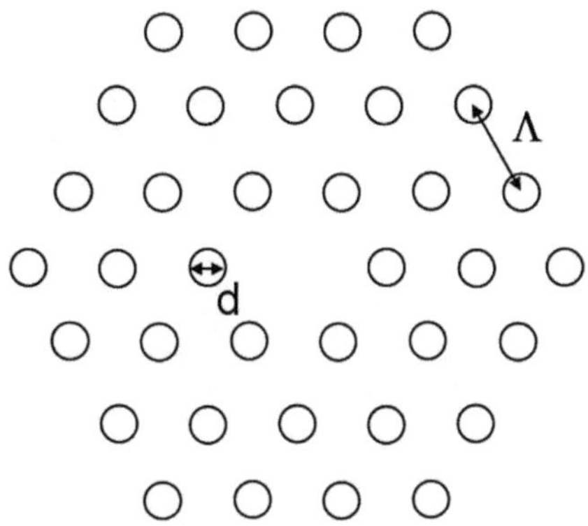


Figure **3.1**: Cross-section of a triangular PCF with the air-hole diameter $d$ and the pitch $\Lambda$ [3.8].

increases, this approximation gets worse, as demonstrated for a holey fiber with a small core and large air-holes, analyzed using the FEM solver [3.16]. Notice that a full-vector analysis is necessary to model PCFs with large air-holes and large index variations and to accurately predict properties, such as dispersion [3.17].

In this study, the design of triangular PCFs has been optimized by properly tailoring the air-hole diameter $d$ and the pitch $\Lambda$, as shown in Fig. 3.1, in order to compensate both the positive dispersion and the positive dispersion slope of single-mode fibers over a wavelength range around 1550 nm. To this aim, triangular PCFs with large air-holes and a small pitch, that is with a small core diameter $d_{\text{core}} = 2\Lambda - d = \Lambda \cdot (2 - d/\Lambda)$, have been considered. In fact, in these conditions the possibility to obtain strong negative dispersion values has been already demonstrated [3.16]. For all the triangular PCFs here studied a proper number of air-hole rings has been considered, in order for the solution to converge toward that of a fiber with an infinite photonic crystal cladding. This results in a considerable reduction of the leakage losses [3.18]. In particular, through the complex FEM formulation, which allows radiation field to be evaluated, as described in Appendix A, it has been shown that, by choosing the ring number between three and nine, leakage losses of fibers with $d/\Lambda$ in the range 0.6–0.9 can be reduced under the Rayleigh scattering limit [3.19, 3.20]. The dispersion parameter $D$ has been derived in the wavelength range 1200 nm–1600 nm.

The first fibers considered have $d/\Lambda = 0.9$ and $\Lambda$ which varies between 0.6 and 1 μm. Figure 3.2a shows their dispersion parameter $D$ for the wavelengths

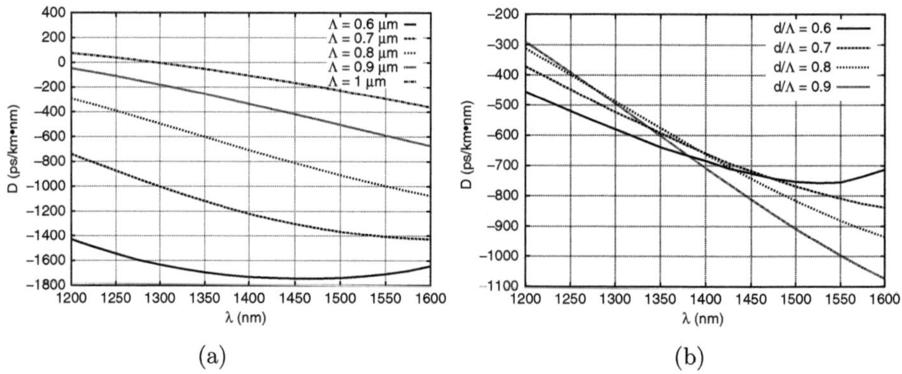


Figure **3.2**: Dispersion parameter for PCFs (a) with $d/\Lambda = 0.9$ and different $\Lambda$ values, and (b) with $\Lambda = 0.8$ μm and different $d/\Lambda$ values [3.8].

between 1200 and 1600 nm. $D$ is always negative if $\Lambda < 1$ μm and becomes positive only for the triangular PCF with $\Lambda = 1$ μm when $\lambda < 1300$ nm. The absolute value of the dispersion parameter increases reducing the hole-to-hole spacing $\Lambda$. For the triangular PCF with $\Lambda = 0.6$ μm $D$ reaches a value around $-1700$ ps/km · nm at 1550 nm, while for conventional DCFs it is typically $-100$ ps/km · nm at this wavelength [3.15,3.21]. The dispersion slope is always negative in the wavelength range considered if $\Lambda \geq 0.7$ μm, while for the PCF with the smallest pitch, $\Lambda = 0.6$ μm, $D$ reaches a minimum at 1475 nm and then the dispersion slope becomes positive.

In order to understand how to optimize the PCF design, the effect of $d$ variation has been also investigated. For this reason the pitch has been fixed to $\Lambda = 0.8$ μm, that is, a middle value between those previously considered, and the ratio $d/\Lambda$ has been varied from 0.9 to 0.6. As shown in Fig. 3.2b, $D$ is always negative in the wavelength range chosen for all the $d/\Lambda$ values. As $d/\Lambda$ decreases from the initial value of 0.9, the dispersion slope changes and becomes positive for the PCF with $d/\Lambda = 0.6$ if $\lambda > 1525$ nm. The minimum value of $D$ at 1550 nm, around $-1000$ ps/km · nm, has been obtained with the largest air-holes, that is, with $d/\Lambda = 0.9$.

Results reported so far are summarized in Fig. 3.3, which shows the dispersion parameter values at 1550 nm. Notice that the dispersion value increases significantly with $\Lambda$ when $d/\Lambda$ is fixed to 0.9, while it slowly decreases when the air-holes become larger, as in the case $\Lambda = 0.8$ μm. This result suggests important technological considerations. In fact, proper pitch values, rather

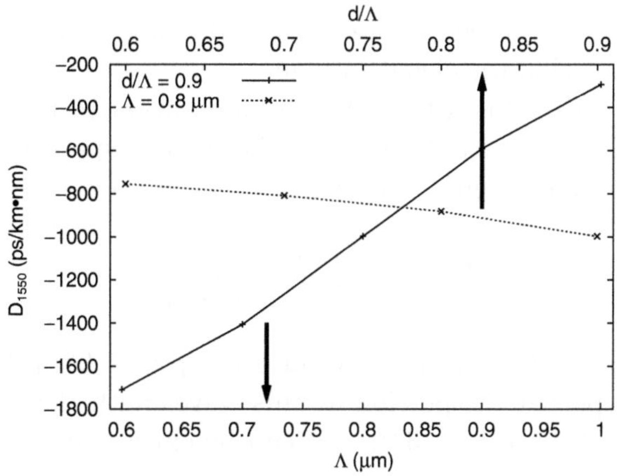


Figure **3.3**: Chromatic dispersion value at 1550 nm for the different triangular PCFs considered [3.9].

than high air-filling fractions, allow to get fibers with dispersion values slightly affected by small variations of the air-hole diameter, eventually introduced by the fabrication process.

The anomalous dispersion of an SMF at 1550 nm is completely compensated by a DCF if

$$D_{\mathrm{SMF}} \cdot L_{\mathrm{SMF}} + D_{\mathrm{DCF}} \cdot L_{\mathrm{DCF}} = 0 \;, \tag{3.1}$$

where $D_{\mathrm{SMF}}$, $D_{\mathrm{DCF}}$, $L_{\mathrm{SMF}}$, and $L_{\mathrm{DCF}}$ are, respectively, the dispersion parameters and the lengths of the single-mode and the dispersion-compensating fibers. For a given SMF, if the absolute value of $D_{\mathrm{DCF}}$ is bigger, the length of the DCF can be shorter. The triangular PCF with $\Lambda = 0.6$ μm and $d/\Lambda = 0.9$, which has the largest value of negative dispersion at 1550 nm, as shown in Fig. 3.2a, can be about 17 times shorter than a classical DCF. Unfortunately this fiber has a positive dispersion curve slope in the third window.

In fact, the dispersion slope is very important, being the parameter which characterizes the dispersion compensation over a wavelength range. In an SMF the slope of $D(\lambda)$ at 1550 nm is positive. The two PCFs, with $\Lambda = 0.6$ μm and $d/\Lambda = 0.9$ in Fig. 3.2a, and with $\Lambda = 0.8$ μm and $d/\Lambda = 0.6$ in Fig. 3.2b, have a positive dispersion slope too, so they are suitable for dispersion compensation only at one wavelength. In particular, the latter PCF has a lower value of $D$ at 1550 nm, $-755$ ps/km · nm. All the other PCFs present a negative dispersion slope at 1550 nm and can be exploited to compensate the anomalous dispersion of an SMF over a wide wavelength range.

In order to verify this aspect, the compensation ratio CR has been calculated [3.15]. CR($\lambda$) is the fraction of the SMF dispersion which the DCF compensates at a wavelength $\lambda$, that is,

$$\mathrm{CR}(\lambda) = \left| \frac{D_{\mathrm{SMF}}(\lambda) \cdot L_{\mathrm{SMF}}}{D_{\mathrm{DCF}}(\lambda) \cdot L_{\mathrm{DCF}}} \right| . \tag{3.2}$$

The value of CR at 1550 nm is 1, because $L_{\mathrm{SMF}}$ and $L_{\mathrm{DCF}}$ have been chosen to perfectly compensate dispersion at this wavelength through Eq. (3.1). By substituting Eq. (3.1) in Eq. (3.2), CR can be expressed as

$$\mathrm{CR}(\lambda) = \frac{D_{\mathrm{SMF}}(\lambda)}{D_{\mathrm{SMF}}} \cdot \frac{D_{\mathrm{DCF}}}{D_{\mathrm{DCF}}(\lambda)} . \tag{3.3}$$

As an example of a standard SMF, the Corning® SMF-28™ has been considered. Its $D(\lambda)$ values have been calculated on a wavelength range of 100 nm through

$$D(\lambda) \approx \frac{S_0}{4} \left[ \lambda - \frac{\lambda_0^4}{\lambda^3} \right] , \tag{3.4}$$

considering a zero-dispersion wavelength $\lambda_0$ of 1311.5 nm and a zero-dispersion slope $S_0$ of 0.092 ps/km $\cdot$ nm$^2$ [3.22]. CR($\lambda$) evaluated for several triangular PCFs is shown on a 100 nm wavelength range for $d/\Lambda = 0.9$ and different $\Lambda$ values in Fig. 3.4a, and for $\Lambda = 0.8$ μm and different $d/\Lambda$ values in Fig. 3.4b. The best compensation can be obtained with the PCF with $d/\Lambda = 0.9$ and $\Lambda = 0.9$ μm, because CR is 0.966 at 1500 nm and 1.016 at 1600 nm. In these cases the residual dispersion is, respectively, $-0.505$ ps/km $\cdot$ nm and 0.318 ps/km $\cdot$ nm. This PCF has $D_{\mathrm{DCF}} = -590$ ps/km $\cdot$ nm, so that, from Eq. (3.1), $L_{\mathrm{DCF}}$ can be only 2.94% of $L_{\mathrm{SMF}}$ to completely compensate the anomalous dispersion at 1550 nm.

Considering the slope of the curves in Fig. 3.4a, an even more flattened CR($\lambda$) over the 100 nm range can be obtained for a triangular PCF with $d/\Lambda = 0.9$ and $\Lambda$ between 0.9 and 1 μm. The dispersion curves for these fibers are shown in Fig. 3.5a, while the corresponding CR values for the wavelengths between 1500 and 1600 nm are reported in Fig. 3.5b. As expected, the optimum pitch value, which provides the best dispersion compensation in the considered wavelength range, is 0.94 μm. For this PCF, that is the one with

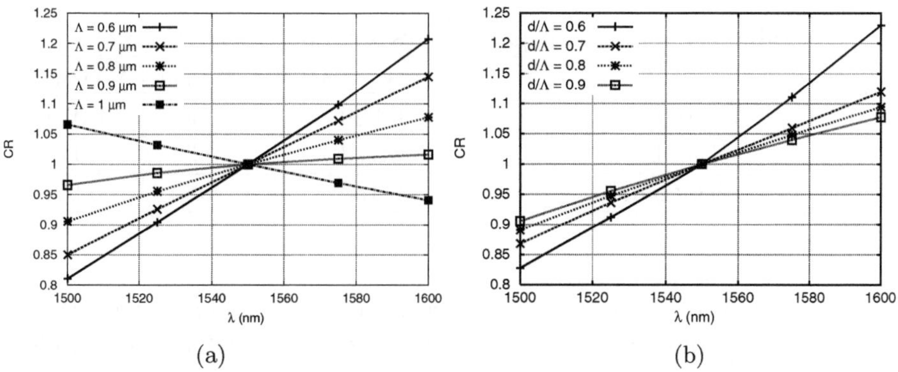


Figure **3.4**: Compensation ratio for PCFs (a) with $d/\Lambda = 0.9$ and different $\Lambda$ values, and (b) with $\Lambda = 0.8$ μm and different $d/\Lambda$ values compensating SMF-28 [3.8].

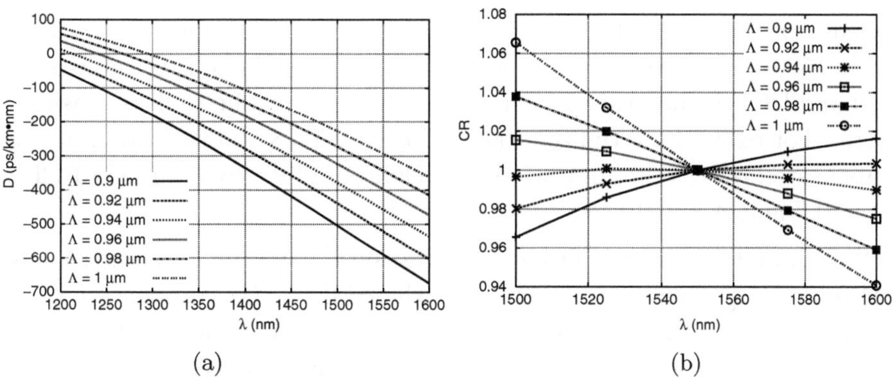


Figure **3.5**: (a) Dispersion parameter and (b) compensation ratio for PCFs with $d/\Lambda = 0.9$ and different $\Lambda$ in the range between 0.9 and 1 μm.

$d/\Lambda = 0.9$ and $\Lambda = 0.94$ μm, the dispersion parameter is $-456$ ps/km · nm at 1550 nm and the residual dispersion is $-0.05$ ps/km · nm and $-0.21$ ps/km · nm at 1500 nm and 1600 nm, respectively.

Moreover, it is important to point out that there is a trade-off between the DCF length and the wavelength range where the dispersion is well compensated. While the PCF with $d/\Lambda = 0.9$ and $\Lambda = 0.6$ μm is the best for the first aspect, the PCF with $d/\Lambda = 0.9$ and $\Lambda = 0.9$ μm is better for WDM systems.

Table 3.1: Important parameters for different DCFs [3.8].

| DCF | $L_{DCF}/L_{SMF}$ | $CR_{1500}$ | $CR_{1600}$ |
|---|---|---|---|
| PCF ($d/\Lambda = 0.9$, $\Lambda = 0.9$ μm) | 0.029 | 0.966 | 1.016 |
| Standard | 0.173 | 0.927 | 1.047 |
| Wideband | 0.183 | 0.9985 | 0.99 |

The behavior of the latter PCF as a dispersion compensating fiber for an SMF-28 can be compared with the one of a standard DCF and a wideband DCF. Typical values of the dispersion and of the dispersion slope at 1550 nm have been chosen [3.21], such as $D_{1550} = -100$ ps/km · nm and $S_{1550} = -0.22$ ps/km · nm$^2$ for a standard DCF, and $D_{1550} = -95$ ps/km · nm and $S_{1550} = -0.33$ ps/km · nm$^2$ for a wideband DCF. A linear behavior has been supposed in the wavelength range between 1500 and 1600 nm. In order to evaluate the efficiency of the different DCFs in a WDM system, the values of $L_{DCF}$ and of CR at 1500 and 1600 nm have been considered and listed in Table 3.1. The wideband DCF is the best over a range of 100 nm, but the PCF is considerably shorter. For example, a typical 100 km long SMF-28 transmission link can be compensated by 18.3 km of a wideband DCF and by only 3 km of a PCF with $d/\Lambda = 0.9$ and $\Lambda = 0.9$ μm.

It is important to underline that the best value of $d/\Lambda$ and $\Lambda$ for the triangular PCF obviously depends on the SMF to be compensated. As an example, consider the Ritekom G-655 fiber, a nonzero dispersion fiber (NZDF) with a dispersion parameter of 8.2 ps/km · nm and a dispersion slope of 0.043 ps/km · nm$^2$ at 1550 nm [3.23]. As shown in Fig. 3.6a and b, the triangular PCF with $d/\Lambda = 0.9$ and $\Lambda = 1$ μm can be assumed as the best DCF. In fact, CR is 0.913 at 1500 nm and 1.047 at 1600 nm, resulting in a residual dispersion of $-0.55$ ps/km · nm and 0.48 ps/km · nm, respectively. Moreover, to completely compensate the anomalous dispersion, the length of the triangular PCF can be 2.8% the length of the NZDF, being $D_{DCF} = -293$ ps/km · nm.

A second important consequence of the small core diameter of the proposed PCFs is the increase of the nonlinear coefficient. While this aspect can be successfully exploited, for example, in Raman amplification, it can be critical for applications like dispersion compensation. In order to evaluate the fiber nonlinearity, the effective area $A_{\text{eff}}$ has been accurately calculated according to Eq. (A.7), since it is inversely related to the nonlinear coefficient. Results

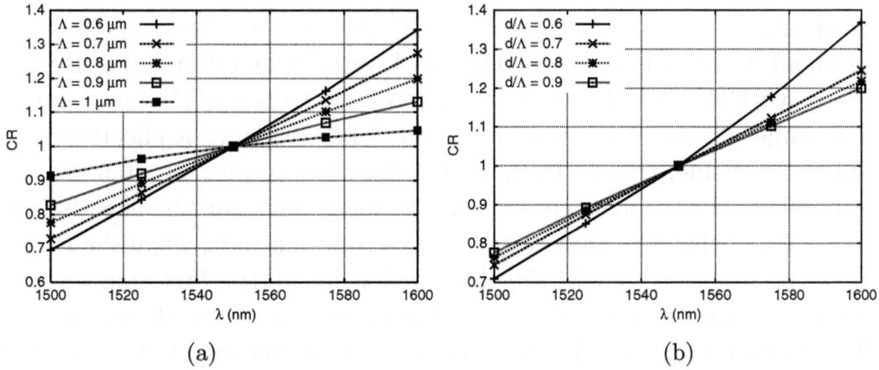


Figure **3.6**: Compensation ratio for PCFs (a) with $d/\Lambda = 0.9$ and different $\Lambda$ values, and (b) with $\Lambda = 0.8$ μm and different $d/\Lambda$ values compensating Ritekom G-655 fiber [3.8].

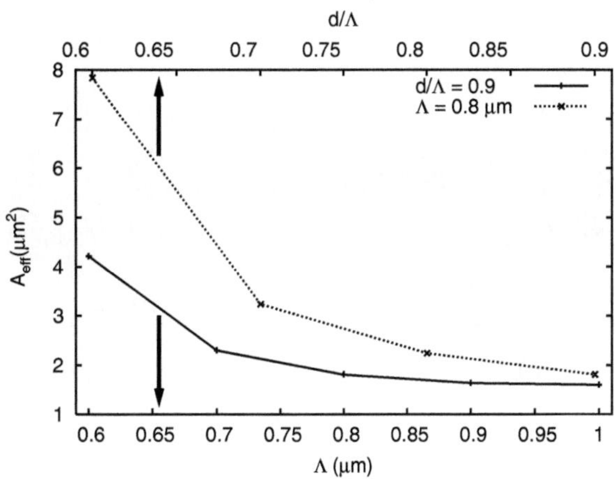


Figure **3.7**: Effective area at 1550 nm for different triangular PCFs with large air-holes and a small pitch [3.9].

are reported in Fig. 3.7. In conclusion, it is important to highlight that all the triangular PCFs considered in the present analysis have a small core diameter, about 1 μm, which can results in large coupling losses with standard fibers [3.24, 3.25]. However, taper holey fiber structures used as a spot-size converter have been recently demonstrated [3.26], providing only 0.3 dB coupling loss with a standard single-mode fiber. For the PCFs with $\Lambda$ fixed to 0.8 μm, $d_{\text{core}}$

becomes smaller when $d/\Lambda$ increases and the $A_{\mathrm{eff}}$ has the same behaviour, being about 8 $\mu m^2$ when $d/\Lambda = 0.6$, and 1.8 $\mu m^2$ when $d/\Lambda = 0.9$. Regarding the PCFs with $d/\Lambda$ fixed to 0.9, it is interesting to notice that $d_{\mathrm{core}}$ increases with $\Lambda$, while $A_{\mathrm{eff}}$ becomes smaller. In fact, for $\Lambda = 1$ μm the effective area has the minimum value, that is 1.6 $\mu m^2$. On the contrary, for $\Lambda = 0.6$ μm, which corresponds to the smallest core diameter, about 0.66 μm, the effective area is 4.2 $\mu m^2$. This behavior can be explained considering that, when the PCF core diameter becomes too small, the silica region inside the first ring, in spite of the large surrounding air-holes, is unable to confine the field, which expands itself on a broader area. This effect is confirmed by looking at the fundamental component of the magnetic field at 1550 nm, shown in Fig. 3.8a and b for two PCFs with $\Lambda = 0.8$ μm, with $d/\Lambda = 0.6$ and $d/\Lambda = 0.9$, respectively, and in Fig. 3.8c and d for two PCFs with $d/\Lambda = 0.9$, with $\Lambda = 0.6$ μm and $\Lambda = 1$ μm, respectively.

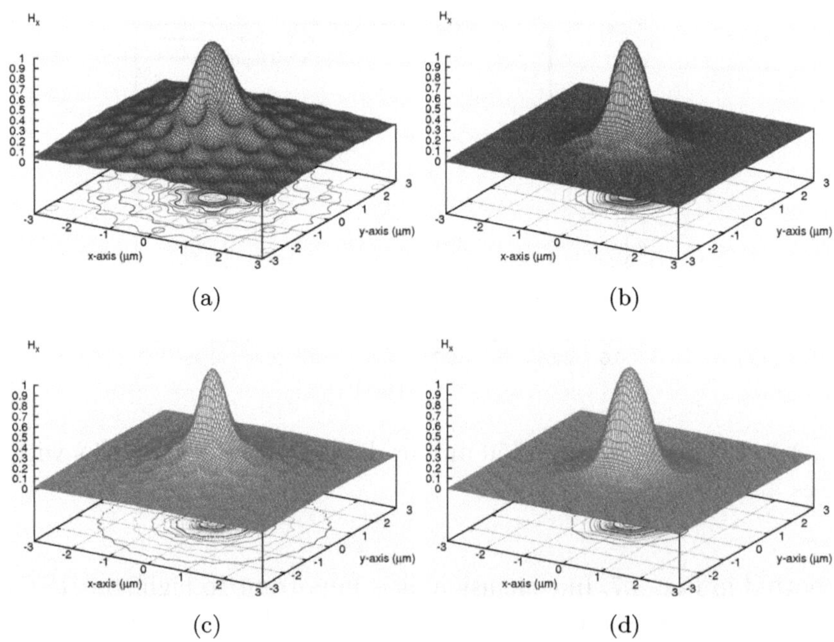


Figure **3.8**: Magnetic field fundamental component at 1550 nm for the two PCFs with $\Lambda = 0.8$ μm and (a) $d/\Lambda = 0.6$, and (b) $d/\Lambda = 0.9$, and for the two PCFs with $d/\Lambda = 0.9$ and (c) $\Lambda = 0.6$ μm, and (d) $\Lambda = 1$ μm.

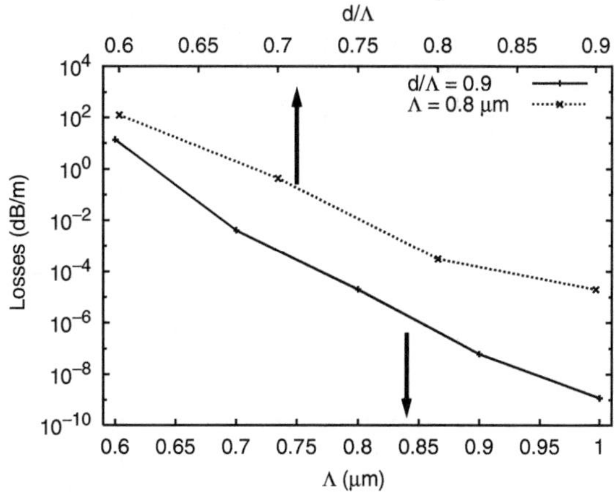


Figure **3.9**: Leakage losses at 1550 nm for different triangular PCFs with large air-holes and a small pitch [3.9].

Last considerations suggest to analyze the losses of these triangular PCFs in order to check the amount of leakage, which can represent a problem for their successful application. Looking at the values evaluated at 1550 nm, reported in Fig. 3.9 for the same fibers, it is possible to notice that, fixed $d/\Lambda$ to 0.9, the leakage losses decrease as soon as the pitch $\Lambda$ varies from 0.6 to 1 μm. In fact, these losses are about 14 dB/m when $\Lambda = 0.6$ μm, while they can be neglected if $\Lambda \geq 0.8$ μm, being lower than $10^{-4}$ dB/m, that is under the Rayleigh scattering limit. Moreover, when $\Lambda$ is fixed to 0.8 μm, the leakage losses decrease as the air-holes become larger, that is for increasing $d/\Lambda$ values. In fact, the guided mode is more confined in the PCFs, due to the higher air-filling fraction [3.19]. Finally, notice that, by increasing the number of air-hole rings in the fiber cross-section, leakage losses can be neglected, being under the Rayleigh limit, also for different triangular PCFs. For example, nine rings are enough for $d/\Lambda > 0.8$ and $\Lambda \geq 0.8$ μm.

## 3.2 Dispersion of square-lattice PCFs

With the previous thorough analysis it has been shown that triangular PCFs with a silica core can be successfully used to compensate the positive dispersion parameter and the dispersion slope of a SMF [3.8]. In order to investigate the

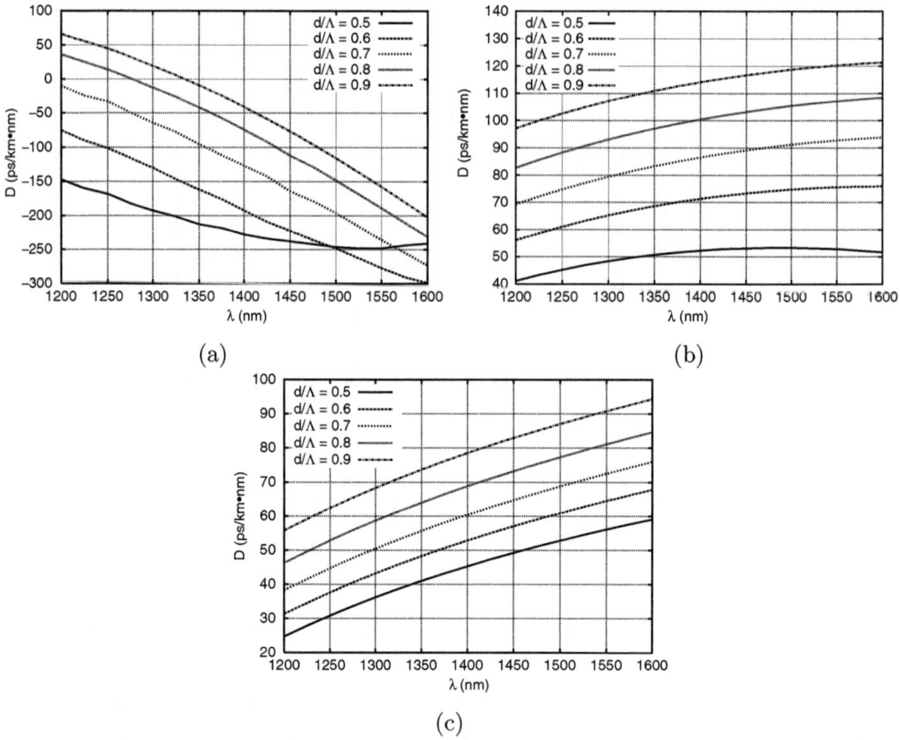


Figure **3.10**: Dispersion curves of the square-lattice PCFs with (a) $\Lambda = 1$ µm, (b) $\Lambda = 2$ µm and (c) $\Lambda = 3$ µm for different $d/\Lambda$ values in the range 0.5–0.9 [3.11].

possibility to design square-lattice fibers, shown in Fig. 2.1a, with the same dispersion characteristics, the dispersion curves of these PCFs with hole-to-hole spacing in the range 1–3 µm, and $d/\Lambda$ between 0.5 and 0.9, previously described in Chapter 2, have been accurately calculated.

Figure 3.10 shows the dispersion curves $D(\lambda)$ of the square-lattice PCFs with different $d/\Lambda$ values and $\Lambda = 1$, 2 and 3 µm, which have been derived by applying the simple finite difference formula of Eq. (A.4) to the effective index values reported in Fig. 2.2 for the wavelengths between 1200 and 1600 nm. Looking at Fig. 3.10a, notice that all the square-lattice PCFs with the smallest pitch, that is 1 µm, have negative dispersion parameter in the C band, around 1550 nm, since the core dimension is very small and the waveguide dispersion dominates on the material one [3.6, 3.27]. The minimum dispersion value at 1550 nm, $-277$ ps/km·nm, has been obtained with the PCF characterized by $\Lambda = 1$ μm and $d/\Lambda = 0.6$. It is important to underline that the $D$ values increase with the air-hole diameter, so only the PCFs with $d/\Lambda \leq 0.7$, that is with small air-holes, have negative dispersion parameter in all the wavelength range here considered. The fiber with the smallest air-holes, that is the one with $d/\Lambda = 0.5$, has a dispersion curve with a minimum, about $-248$ ps/km·nm, around 1550 nm and a positive dispersion slope for the longer wavelengths. The other square-lattice PCFs, with $d/\Lambda \geq 0.6$, have negative dispersion slope, so they can be used as dispersion compensating fibers. As it has been already demonstrated for the triangular PCFs [3.6, 3.27], the influence of the waveguide dispersion decreases when the pitch becomes larger. This is confirmed also in Fig. 3.10b and c, which show results for $\Lambda$ equal to 2 and 3 μm, respectively. Notice that the dispersion parameter of all these PCFs is positive, independently from the air-hole dimension, that is from the $d/\Lambda$ value. It is interesting to underline that, as the pitch increases for a fixed $d/\Lambda$ value, the dispersion slope of the curves becomes more positive. Moreover, a change of $d/\Lambda$ causes a small difference in the dispersion parameter values, of about 8 ps/km·nm in all the considered wavelength range, for the PCFs with the higher $\Lambda$, that is 3 μm. Notice that the dispersion curve of the square-lattice PCF with $d/\Lambda = 0.5$ and $\Lambda = 2$ μm is quite flat, around the value of 53 ps/km·nm, from 1425 to 1550 nm, as shown in Fig. 3.10b.

Figure 3.11 allows to understand how the dispersion properties of the square-lattice PCFs change as a function of the pitch $\Lambda$ for a fixed $d/\Lambda$ value. Fibers with $d/\Lambda = 0.9$ have been considered, whose $n_{\text{eff}}$ values are reported in Chapter 2 in Fig. 2.3. Notice that an increase of 0.5 μm in the pitch value, that is from 1 to 1.5 μm, causes a significant change in the dispersion curve. In fact, there is a great difference between the dispersion parameter values of the two PCFs, which increases with the wavelength, being about 56 ps/km·nm at 1250 nm and about 310 ps/km·nm at 1600 nm. Moreover, the dispersion slope, which is negative for the PCF with $\Lambda = 1$ μm in all the wavelength range considered, becomes almost null in the wavelength range between 1200 and 1450 nm, and positive for the longer wavelengths for the PCF with $\Lambda = 1.5$ μm. When $\Lambda \geq 2$ μm, the slope of the dispersion curves is always positive. Finally, the dispersion parameter values, which are all higher than 50 ps/km·nm for these PCFs, decrease as the pitch $\Lambda$ increases from 2 to 3 μm.

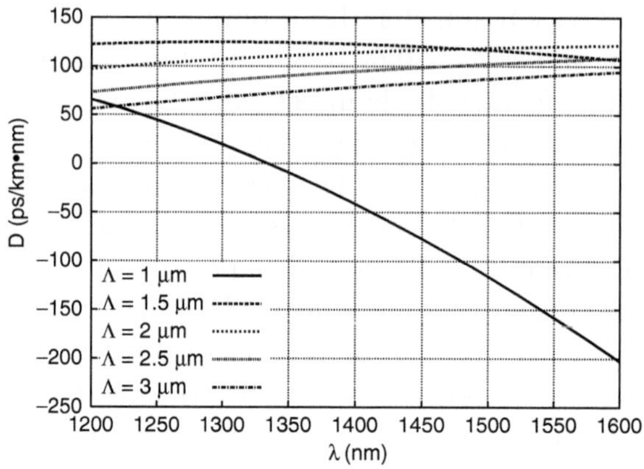


Figure **3.11**: Dispersion curves of the square-lattice PCFs with $d/\Lambda = 0.9$ for different $\Lambda$ values between 1 and 3 μm [3.11].

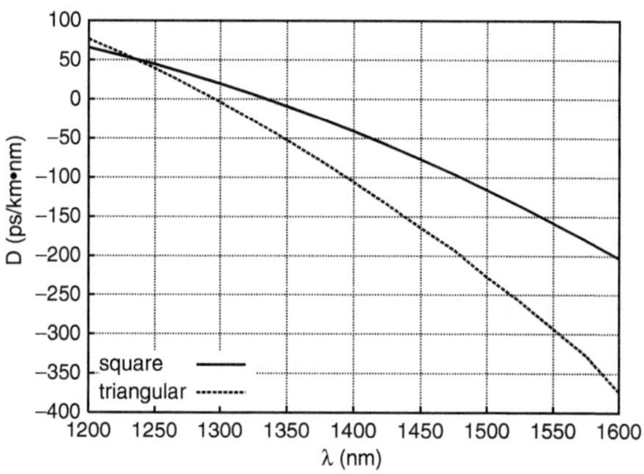


Figure **3.12**: Comparison of the dispersion parameter for the square-lattice PCF and the triangular one with $d/\Lambda = 0.9$ and $\Lambda = 1$ μm [3.11].

After the comparison of the guiding properties described in Chapter 2, the dispersion curves for the square-lattice PCF and the triangular one with $\Lambda = 1$ μm and $d/\Lambda = 0.9$ have been accurately evaluated. It has been already demonstrated [3.8] that the triangular PCF with these geometric parameters has negative dispersion and dispersion slope, as shown in Fig. 3.2a, and can be successfully used as a dispersion compensating fiber for a NZDF, that is the Ritekom G-655 fiber, as reported in Fig. 3.6a. As it is shown in Fig. 3.12, both the fibers have negative dispersion, which is greater in module for the triangular PCF. For example, at 1550 nm $D = -293$ ps/km · nm for the triangular PCF and $D = -157$ ps/km · nm for the square-lattice one. The last PCF could compensate the positive dispersion of the NZDF in a wider wavelength range, since its dispersion slope is lower around 1550 nm. In fact, its CR values at 1500 nm and 1600 nm, being 0.964 and 1.007, respectively, are closer to the optimum value, that is 1, than those for the triangular PCF, which are 0.913 at 1500 nm and 1.047 at 1600 nm [3.8]. However, a square-lattice PCF longer than the triangular one would be necessary to completely compensate the dispersion of the NZDF at 1550 nm, due to its lower negative dispersion parameter value at this wavelength.

A final analysis of the properties of the square-lattice and the triangular PCFs is reported in Fig. 3.13 for different values of the hole-to-hole spacing. A small $d/\Lambda$ value, that is 0.5, has been chosen for the comparison, so that the triangular PCF is single-mode in all the wavelength range considered also for the largest pitch $\Lambda = 3$ μm [3.28, 3.29]. It is interesting to notice that the square-lattice PCF has a higher dispersion parameter than the triangular one when the pitch is small, that is 1 μm, and lower $D$ values when the hole-to-hole distance is large, that is $\Lambda = 3$ μm, as reported in Fig. 3.13. The dispersion slope is only slightly influenced by the geometric characteristics of

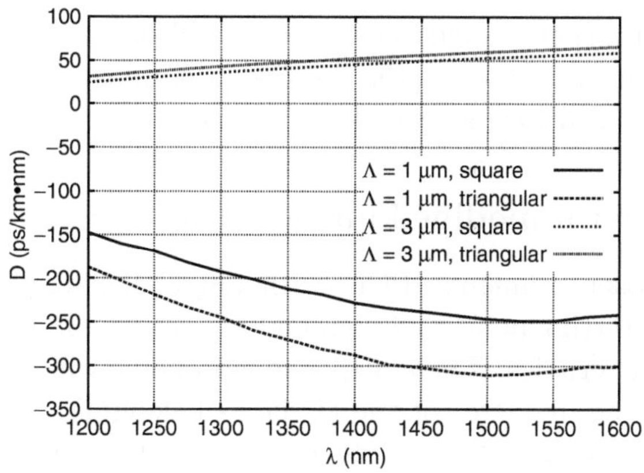


Figure **3.13**: Comparison of the dispersion parameter values for the square-lattice PCFs and the triangular ones with $d/\Lambda = 0.5$, for $\Lambda = 1$ and 3 μm [3.11].

the lattice, being similar for the two PCFs. The comparison between the two PCFs which involves the effective area values in the wavelength range between 1200 and 1600 nm has been already described in Chapter 2.

## 3.3 Dispersion-flattened triangular PCFs

Results reported earlier in this chapter have proved that the PCF dispersion properties can be engineered by changing the geometric parameters, that is $\Lambda$ and $d$. For example, it has been demonstrated that, in order to design triangular PCFs for dispersion compensation, it is necessary to choose large air-holes and small pitch values. Now it is interesting to investigate how the geometric parameters of the PCF cross-section can be changed to obtain fibers with a flat dispersion curve and the zero-dispersion wavelength around 1550 nm. Notice that PCFs with these characteristics and with a small effective area, that is a high nonlinear coefficient, are suitable for a great number of telecommunication applications, such as wavelength conversion [3.30] or optical parametric amplification [3.31].

In literature dispersion-flattened triangular PCFs have been obtained, for example, by keeping fixed the geometry of the first ring of air-holes around the core and by progressively enlarging the holes of the outer rings [3.32], or by introducing dopants in the fiber cross-section center to realize a hybrid core region with a threefold symmetry [3.33]. In the analysis here reported two different approaches have been proposed in order to design triangular PCFs with a flat dispersion curve and the zero-dispersion wavelength in the C band. Notice that in both cases the dispersion parameter $D$ has been derived according to Eq. (A.4), as described in Appendix A.

### 3.3.1 PCFs with modified air-hole rings

In the first approach, triangular PCFs with a high air-filling fraction have been considered as a starting point to design highly nonlinear fibers with the desired dispersion characteristics. Their dispersion properties have been studied by modifying only the diameter of the air-holes belonging to first, second, and third ring. In particular, the present analysis starts from large air-holes and small pitch PCFs, which can successfully compensate both the positive dispersion and the positive dispersion slope [3.8], assuring, at the same time, small effective area and thus high nonlinear coefficient. Moreover, the attention

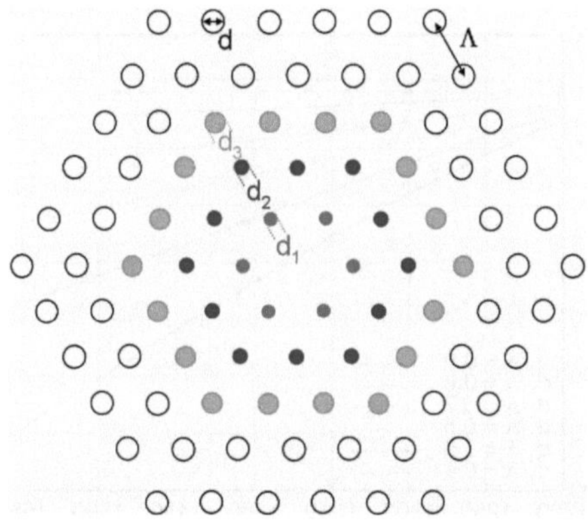


Figure **3.14**: Cross-section of the triangular PCF considered: $d_1$, $d_2$, and $d_3$ are the air-hole diameters in the first, second, and third ring, respectively [3.12].

posed on the geometry of the first air-hole rings can provide a further insight on their role, allowing to separately evaluate their effect.

The present study starts from the triangular PCF with $d/\Lambda = 0.9$ and $\Lambda = 0.9$ μm. As shown in Fig. 3.14, the air-hole diameters of the first, second, and third ring are, respectively, $d_1$, $d_2$, and $d_3$. The dispersion properties of this PCF, that is a negative dispersion parameter and a negative dispersion slope between 1200 and 1600 nm, as reported in Fig. 3.2a, can be modified by changing the air-hole diameter in the first three rings, without significantly affecting its good nonlinear characteristics. In fact, it has been evaluated, according to Eq. (A.7), that the effective area is about 1.6 $\mu m^2$ at 1550 nm, so its nonlinear coefficient value is high, about 65 $(W{\cdot}km)^{-1}$. It is important to underline that, considering nine air-hole rings in the PCF cross-section, the leakage losses at 1550 nm can be neglected, being under the Rayleigh scattering limit [3.20].

Initially, only the air-hole diameter $d_1$ in the first ring has been changed, while all the other geometric characteristics of the PCF, that is the pitch $\Lambda$ and the number of the air-hole rings, have been kept constant. As shown in Fig. 3.15, by reducing $d_1/\Lambda$ to 0.8, 0.7, 0.6, 0.5, and 0.4, the dispersion parameter increases, as well as the dispersion slope, for all the wavelengths between 1200 and 1600 nm. Notice that for the PCF with $d_1/\Lambda = 0.4$ the dispersion slope becomes positive and the dispersion parameter at 1550 nm increases to 24 ps/km · nm. The air-holes belonging to the first ring, which surround the

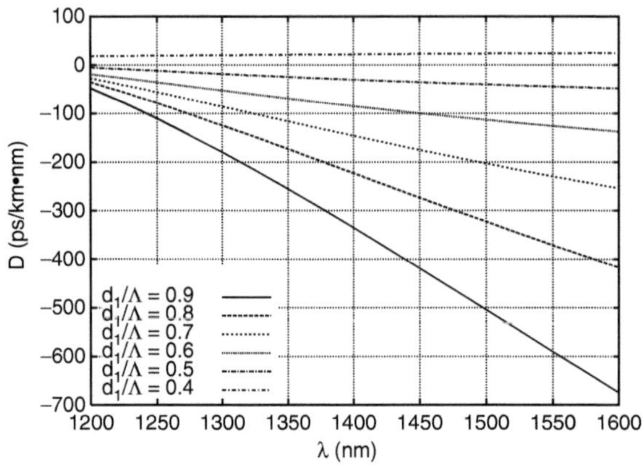


Figure **3.15**: Dispersion parameter of the PCF with $d/\Lambda = 0.9$ and $\Lambda = 0.9$ μm for different $d_1/\Lambda$ values [3.12].

silica core, have a strong influence on the PCF dispersion properties, since the guided-mode field is strictly confined in the central region of the large-hole PCF cross-section. Although the air-hole diameter decreases, the effective area at 1550 nm is not significantly modified, since it becomes 2.8 $\mu m^2$ for the PCF with $d_1/\Lambda = 0.4$, that is less than twice the value for the PCF with $d_1/\Lambda = d/\Lambda = 0.9$.

Bringing back $d_1/\Lambda$ to the original value, that is 0.9, and then decreasing only the air-hole diameter in the second ring, the dispersion parameter becomes more negative in all the wavelength range considered, as reported in Fig. 3.16 for $d_2/\Lambda = 0.8$, 0.7, and 0.6. It is interesting to notice that the PCFs with $d_2/\Lambda$ in the range 0.7–0.9 have also a negative dispersion slope, so they can be successfully used as dispersion compensating fibers. The most negative dispersion value, $-1426$ ps/km · nm, has been obtained at 1500 nm for the PCF with $d_2/\Lambda = 0.6$. However, the slope of this PCF dispersion curve is positive around 1550 nm. Finally, it is important to highlight that the decrease of $d_2$ produces a wider silica region between the first and the third air-hole rings, so that the guided-mode field is less confined in the PCF silica core and the effective area increases to 3.5 $\mu m^2$.

The influence of the air-hole diameter of the third ring, that is of $d_3$, on the PCF dispersion properties is demonstrated by the results shown in Fig. 3.17. If $d_3/\Lambda$ decreases to 0.8, 0.7, and 0.6, the dispersion parameter is not significantly modified at the shorter wavelengths, due to the tight confinement of the

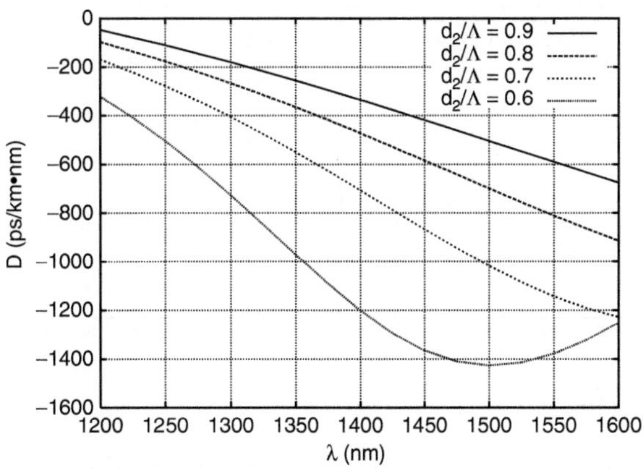


Figure **3.16**: Dispersion parameter of the PCF with $d/\Lambda = 0.9$ and $\Lambda = 0.9$ μm for different $d_2/\Lambda$ values [3.12].

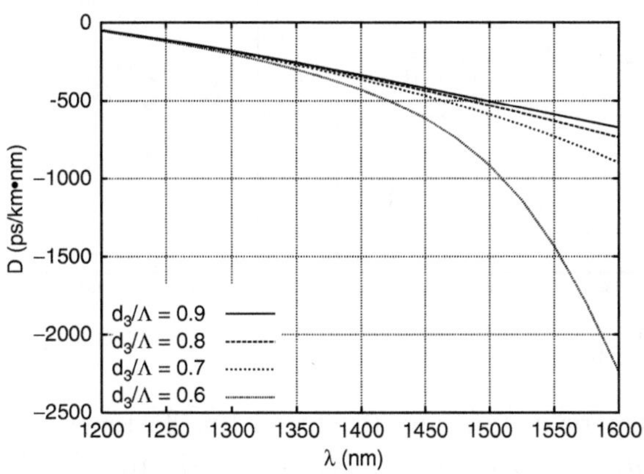


Figure **3.17**: Dispersion parameter of the PCF with $d/\Lambda = 0.9$ and $\Lambda = 0.9$ μm for different $d_3/\Lambda$ values [3.12].

guided-mode field in the silica core. However, at wavelengths longer than 1400 nm the dispersion parameter value decreases much faster with $d_3/\Lambda$. The PCF with $d_3/\Lambda = 0.6$ has a very high negative dispersion parameter value at 1550 nm, about $-1430$ ps/km · nm, a negative dispersion slope and an effective area of 1.88 μm$^2$.

The previous considerations about the influence of the air-holes of the first three rings are useful to design triangular PCFs with the zero-dispersion

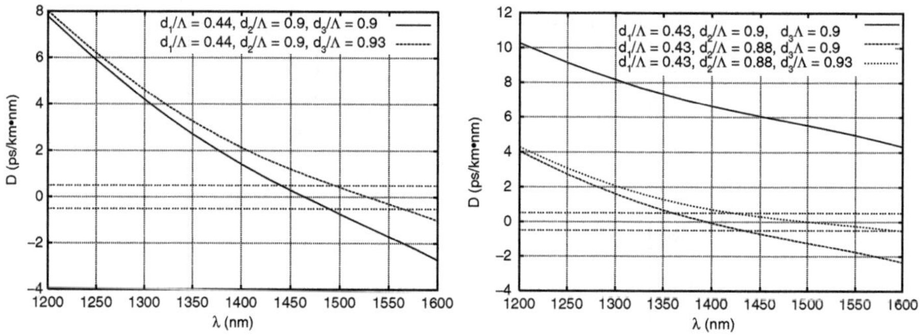


(a) (b)

Figure **3.18**: Dispersion tailoring to obtain a flattened dispersion curve for the PCFs with (a) $d_1/\Lambda = 0.44$ and (b) $d_1/\Lambda = 0.43$ [3.12].

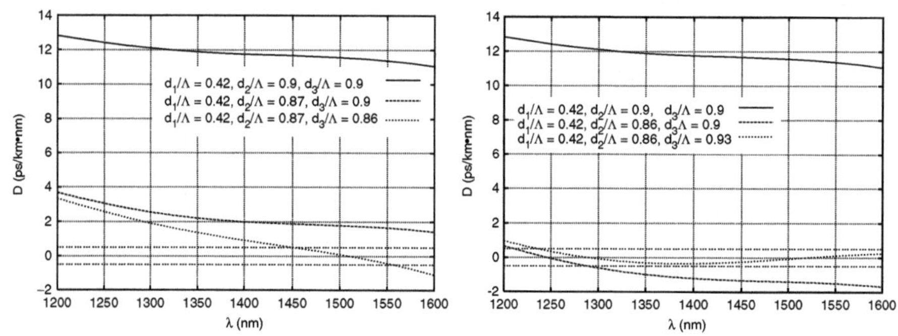


(a) (b)

Figure **3.19**: Dispersion tailoring to obtain the flattened dispersion curve for the PCFs with $d_1/\Lambda = 0.42$ and (a) $d_2/\Lambda = 0.87$, and (b) $d_2/\Lambda = 0.86$ [3.12].

wavelength around 1550 nm and a low-dispersion slope, without significantly increasing the effective area and, as a consequence, without reducing the nonlinear coefficient. A possible procedure is explained in the following and results are shown in Figs. 3.18 and 3.19. The value of $d_1/\Lambda$ is decreased in order to obtain a flat dispersion curve in a wavelength range as large as possible. In this way, the dispersion parameter becomes positive for all the wavelengths considered, so it is necessary to reduce $d_2$ to lower it.

The final dispersion tailoring is made increasing $d_3/\Lambda$, in order to obtain higher dispersion parameter values only at longer wavelengths. Two sets of dispersion curves are reported as examples in Fig. 3.18a and b for two different values of $d_1/\Lambda$, that is 0.44 and 0.43, respectively. Looking at Fig. 3.18a, it is possible to notice that, only by decreasing $d_1/\Lambda$ to 0.44, the dispersion curve presents a zero-dispersion wavelength around 1465 nm, so $d_2$ has been left unchanged. With the choice of $d_3/\Lambda = 0.93$, the zero-crossing of the dispersion curve moves to 1529 nm, but $D$ values remain between $\pm 0.5$ ps/km · nm in a quite small wavelength range, that is about 70 nm. On the contrary, the PCF with $\Lambda = 0.9$ μm, $d_1/\Lambda = 0.43$, $d_2/\Lambda = 0.88$, $d_3/\Lambda = 0.93$ and $d/\Lambda = 0.9$ has dispersion properties much more similar to the desired ones. In fact, as shown in Fig. 3.18b, its dispersion parameter values are between $\pm$ 0.5 ps/km · nm from 1425 to 1600 nm, and the zero-dispersion wavelength is around 1500 nm. Moreover, its effective area is only 2.76 $\mu m^2$ at 1550 nm, which assures a nonlinear coefficient of about 42 $(W{\cdot}km)^{-1}$. A final example of the proposed dispersion tailoring process is reported in Fig. 3.19, where two sets of dispersion curves are shown for the same $d_1/\Lambda$, equal to 0.42, and different $d_2/\Lambda$ values. By choosing $d_2/\Lambda = 0.87$, the dispersion curve results flat around the value of 1.65 ps/km · nm in the wavelength range between 1425 and 1600 nm, so, differently from the previous cases, it is necessary to decrease $d_3/\Lambda$ in order to slightly decrease $D$ values, thus obtaining the zero-dispersion wavelength around 1550 nm. In fact, as shown in Fig. 3.19a, the dispersion parameter values of the PCF with $\Lambda = 0.9$ μm, $d_1/\Lambda = 0.42$, $d_2/\Lambda = 0.87$, $d_3/\Lambda = 0.86$, and $d/\Lambda = 0.9$ are between $\pm 0.5$ ps/km · nm from 1455 to 1560 nm, and its zero-dispersion wavelength is around 1510 nm. Better results have been obtained with a slightly smaller diameter of the air-holes belonging to the second ring, that is $d_2/\Lambda = 0.86$. Looking at Fig. 3.19b, it is possible to notice that the dispersion curve of the PCF with $\Lambda = 0.9$ μm, $d_1/\Lambda = 0.42$, $d_2/\Lambda = 0.86$, $d_3/\Lambda = 0.93$, and $d/\Lambda = 0.9$, whose effective area is 2.84 $\mu m^2$, is ultraflattened, since the dispersion parameter values are between $\pm 0.5$ ps/km · nm in a 375 nm wavelength range.

### 3.3.2 Triangular-core PCFs

Among the highly nonlinear triangular PCFs with flattened dispersion curve and zero-dispersion wavelength around 1550 nm, a novel one with a triangular hybrid core region, obtained by replacing four air-holes with a central germanium up-doped area and three fluorine down-doped regions, has been recently proposed [3.33, 3.34]. Figure 3.20 reports the microscope picture of the realized PCF and a schematic of the materials which constitute the fiber

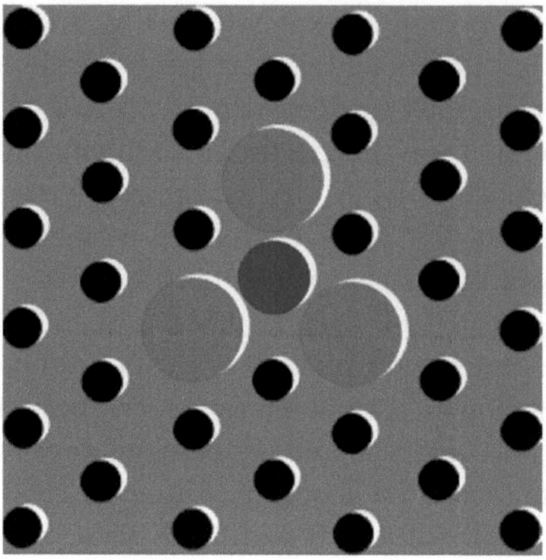

Figure **3.20**: Schematic of the cross-section of the triangular-core fiber presented in [3.33]. The core regions is formed by an up-doped central element surrounded by three down-doped regions and three air-holes.

core. The presence of different dopants in the fiber cross-section offers a further possibility to control the dispersion curve and the nonlinear coefficient. However, this advantage is paid in terms of an increase of the technological effort in the fabrication process.

Starting from these considerations, the second approach followed in the present study consists in designing all-silica triangular-core PCFs with a flattened dispersion curve, the zero-dispersion wavelength around 1550 nm and a high nonlinear coefficient, without the need of adding doped areas in the transverse section. In fact, the PCFs here considered have a silica core with a triangular shape, obtained by removing the central air-hole in the fiber cross-section, and by reducing the diameter of the three surrounding air-holes, which correspond to the fluorine down-doped areas of the fiber proposed in [3.33]. In this way, the possibility to control the refractive index profile by properly changing only the dimension of the air-holes, without the need of any dopants, which is one of the main advantages offered by PCFs, has been exploited. It is important to underline that all the studied PCFs, which are simply made of silica, have a triangular lattice of air-holes in the cross-section, characterized by the pitch $\Lambda$ and the air-hole diameter $d$. Three of the air-holes belonging to the first ring have a different diameter $d_{\mathrm{f}} < d$, as shown in Fig. 3.21, so

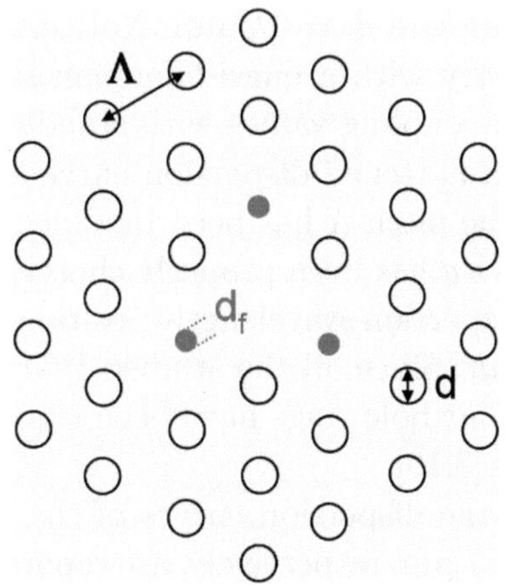


Figure **3.21**: Cross-section of an all-silica triangular-core PCF [3.14].

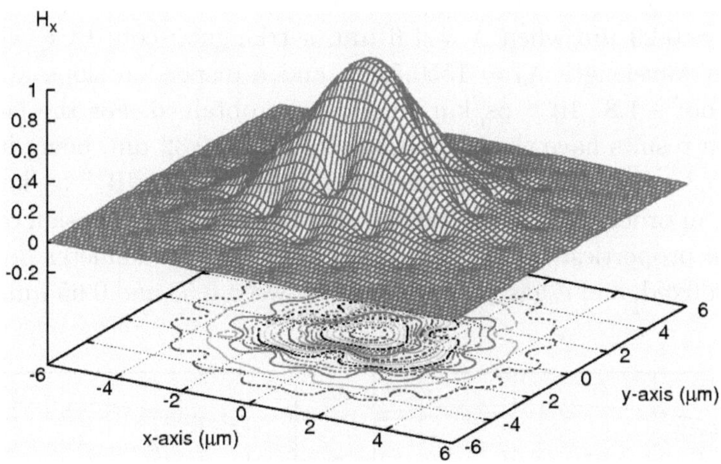


Figure **3.22**: Fundamental component of the guided-mode magnetic field at $\lambda = 1550$ nm for the PCF with $\Lambda = 1.7$ μm, $d = 0.54$ μm and $d_\mathrm{f} = 0.2$ μm [3.14].

that the PCF refractive index profile is similar to the one of the hybrid-core nonlinear fiber in [3.33]. In fact, the core refractive index is higher than the cladding one, while the three smaller air-holes act like the fluorine down-doped areas. The PCF core sustains a guided mode, whose magnetic field fundamental component is reported in Fig. 3.22 for the particular case of $\lambda = 1550$ nm,

$\Lambda = 1.7$ μm, $d = 0.54$ μm and $d_f = 0.2$ μm. Notice that the field distribution has a triangular symmetry with a quasi-Gaussian shape in the center of the core, thus allowing high coupling values with standard fibers [3.33].

In order to achieve a flattened dispersion curve around 1550 nm with the triangular-core PCFs, the pitch $\Lambda$ has been modified in the range 1.4–1.7 μm and the air-hole diameter $d$ has been properly chosen between 0.5 and 0.7 μm. In addition, the zero-dispersion wavelength position in the C band has been optimized by changing $d_f$. Since all the studied PCFs have small $d/\Lambda$ values, between 0.3 and 0.4, 12 air-hole rings have been considered in order to obtain negligible leakage losses [3.19].

In Fig. 3.23a and b the dispersion curves of the PCFs with $d = 0.65$ μm and $\Lambda$ equal to 1.6 and 1.7 μm, respectively, are reported for different $d_f$ values. Notice that for both the considered $\Lambda$ values the $D$ parameter decreases as $d_f$ varies from 0 to 0.3 μm. A further increase of this diameter to 0.4 μm causes a significant, but undesired change in the dispersion curve slope. By properly fixing $d_f = 0.29$ μm when $\Lambda = 1.6$ μm, a triangular-core PCF with a zero-dispersion wavelength $\lambda_0 = 1550.5$ nm and a dispersion slope at $\lambda_0$, called $S_0$, of about $-1.8 \cdot 10^{-2}$ ps/km · nm$^2$ can be obtained. For the larger pitch, simulation results have shown that the best $d_f$ is 0.32 μm, being in this case $\lambda_0 \simeq 1563.3$ nm and the dispersion slope around $-1.3 \cdot 10^{-2}$ ps/km · nm$^2$.

Then, in order to show the influence of the air-hole dimension on the PCF dispersion properties, the pitch $\Lambda = 1.7$ μm and the diameter $d_f = 0.2$ μm have been fixed, and $d$ has been changed between 0.53 and 0.65 μm. As shown

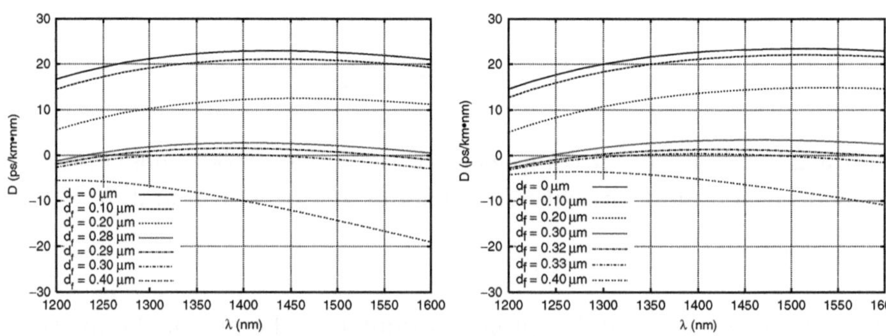


(a) (b)

Figure **3.23**: Dispersion curves as a function of $d_f$ for the PCFs with $d = 0.65$ μm and (a) $\Lambda = 1.6$ μm, and (b) $\Lambda = 1.7$ μm [3.14].

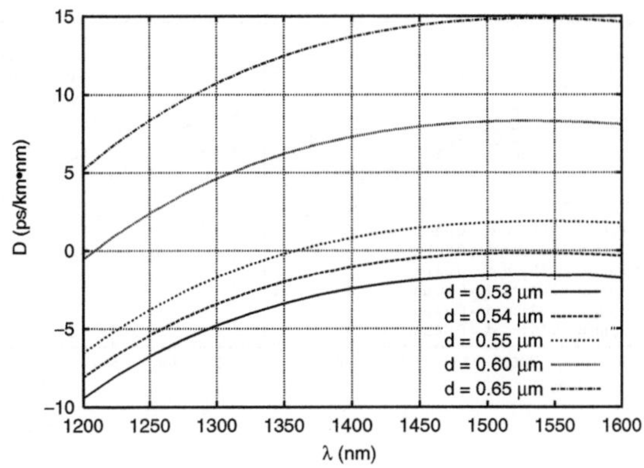


Figure **3.24**: Dispersion curves as a function of $d$ for the PCFs with $d_\text{f} = 0.2$ μm and $\Lambda = 1.7$ μm [3.14].

in Fig. 3.24, $D$ values decrease in all the wavelength range considered when the air-holes become smaller, while the slope of the dispersion curve is only slightly modified. The dispersion parameter is always negative when $d = 0.54$ μm, reaching a maximum of about $-0.14$ ps/km · nm at 1525 nm, with a very low $S_0$, that is about $-1.7 \cdot 10^{-3}$ ps/km · nm$^2$ at 1550 nm. However, the small diameter of all the air-holes in the triangular lattice of this PCF results in a lower field confinement, which limits the value of the nonlinear coefficient.

A similar analysis has been performed also for different configurations of the PCF cross-section. Figure 3.25a and b shows the best dispersion curves obtained considering new $\Lambda$ values with proper air-hole diameters $d$ and $d_\text{f}$. It is important to underline that the PCF with $\Lambda = 1.4$ μm has $d_\text{f} = 0$ μm, that is the three air-holes with diameter $d_\text{f}$ have been completely removed. Also in this case it is possible to achieve a good dispersion slope, about $-3.8 \cdot 10^{-2}$ ps/km · nm$^2$, with a high nonlinear coefficient $\gamma = 10.92$ (W·km)$^{-1}$, calculated according to Eq. (A.8) in Appendix A.

In summary, simulation results have demonstrated that it is possible to design triangular PCFs with flattened dispersion curve, zero-dispersion wavelength around 1550 nm and high nonlinear coefficient with both the proposed approaches, that is by modifying the diameter of the air-holes in the first three rings, as well as by properly choosing the dimension of three air-holes belonging to the first ring around the fiber silica core. Notice that the first triangular PCFs here designed have flatter dispersion curves and smaller effective area

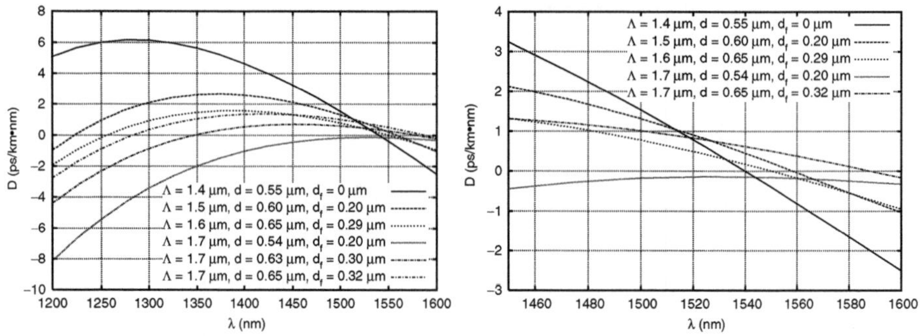


(a) (b)

Figure **3.25**: $D$ versus the wavelength for the best designed triangular-core PCFs in the range (a) 1200–1600 nm and (b) 1450–1600 nm [3.14].

than the triangular-core ones. However, as a drawback, their small pitch, that is 0.9 μm, causes a reduced dimension of the silica core and, as a consequence, an increase of the coupling losses toward standard SMFs, besides some further difficulties in their fabrication process.

# Bibliography


[3.1] P. J. Bennett, T. M. Monro, and D. J. Richardson, "Toward practical holey fiber technology: fabrication, splicing, modeling, and characterization," *Optics Letters*, vol. 24, pp. 1203–1205, Sept. 1999.

[3.2] A. Bjarklev, J. Broeng, K. Dridi, and S. E. Barkou, "Dispersion properties of photonic crystal fibres," in *Proc. European Conference on Optical Communication ECOC 1998*, Sept. 20–24, 1998, pp. 135–136.

[3.3] J. C. Knight, J. Arriaga, T. A. Birks, A. Ortigosa-Blanch, W. J. Wadsworth, and P. St. J. Russell, "Anomalous dispersion in photonic crystal fiber," *IEEE Photonics Technology Letters*, vol. 12, pp. 807–809, July 2000.

[3.4] W. H. Reeves, J. C. Knight, P. St. J. Russell, and P. J. Roberts, "Demonstration of ultra-flattened dispersion in photonic crystal fibers," *Optics Express*, vol. 10, pp. 609–613, July 2002. Available at: http://www.opticsexpress.org/abstract.cfm?URI=OPEX-10-14-609

[3.5] A. Ferrando, E. Silvestre, J. J. Miret, and P. Andrés, "Nearly zero ultra-flattened dispersion in photonic crystal fibers," *Optics Letters*, vol. 25, pp. 790–792, June 2000.

[3.6] A. Ferrando, E. Silvestre, P. Andrés, J. J. Miret, and M. V. Andrés, "Designing the properties of dispersion-flattened photonic crystal fibers," *Optics Express*, vol. 9, pp. 687–697, Dec. 2001. Available at: http://www.opticsexpress.org/abstract.cfm?URI=OPEX-9-13-687

[3.7] G. Renversez, B. Kuhlmey, and R. McPhedran, "Dispersion management with microstructured optical fibers: ultraflattened chromatic dispersion with low losses," *Optics Letters*, vol. 28, pp. 989–991, June 2003.

[3.8] F. Poli, A. Cucinotta, M. Fuochi, S. Selleri, and L. Vincetti, "Characterization of microstructured optical fibers for wideband dispersion compensation," *Journal of Optical Society of America A*, vol. 20, pp. 1958–1962, Oct. 2003.

[3.9] F. Poli, A. Cucinotta, M. Fuochi, S. Selleri, and L. Vincetti, "Dispersion and nonlinear properties of triangular photonic crystal fibers with large air-holes and small pitch," in *Proc. European Conference on Optical Communication ECOC 2003*, Rimini, Italy, Sept. 21–25, 2003.

[3.10] M. Fuochi, F. Poli, S. Selleri, and A. Cucinotta, "Dispersion and dispersion slope compensation through photonic crystal fibers," in *Proc. Progress in Electromagnetics Research Symposium PIERS 2003*, Honolulu, Hawaii, USA, Oct. 13–16, 2003.

[3.11] A. H. Bouk, A. Cucinotta, F. Poli, and S. Selleri, "Dispersion properties of square-lattice photonic crystal fibers," *Optics Express*, vol. 12, pp. 941–946, Mar. 2004. Available at: http://www.opticsexpress.org/abstract.cfm?URI=OPEX-12-5-941

[3.12] F. Poli, A. Cucinotta, S. Selleri, and A. H. Bouk, "Tailoring of flattened dispersion in highly nonlinear photonic crystal fibers," *IEEE Photonics Technology Letters*, vol. 16, pp. 1065–1067, Apr. 2004.

[3.13] F. Poli, F. Adami, M. Foroni, L. Rosa, A. Cucinotta, and S. Selleri, "Optical parametric amplification in all-silica triangular-core photonic crystal fibers," *Applied Physics B*, vol. 81, pp. 251–255, July 2005.

[3.14] S. Selleri, A. Cucinotta, F. Poli, M. Foroni, and L. Rosa, "Optical parametric amplification in dispersion-flattened highly nonlinear photonic crystal fibers," in *Proc. International Congress on Optics and Optoelectronics SPIE-COO 2005*, Warsaw, Poland, Aug. 28–2 Sept. 2005.

[3.15] T. A. Birks, D. Mogilevtsev, J. C. Knight, and P. St. J. Russell, "Dispersion compensation using single-material fibers," *IEEE Photonics Technology Letters*, vol. 11, pp. 674–676, June 1999.

[3.16] A. Cucinotta, S. Selleri, L. Vincetti, and M. Zoboli, "Holey fiber analysis through the finite-element method," *IEEE Photonics Technology Letters*, vol. 14, pp. 1530–1532, Nov. 2002.

[3.17] T. M. Monro, D. J. Richardson, N. G. R. Broderick, and P. J. Bennett, "Modeling large air fraction holey optical fibers," *IEEE/OSA Journal of Lightwave Technology*, vol. 18, pp. 50–56, Jan. 2000.

[3.18] T. P. White, R. C. McPhedran, C. M. de Sterke, L. C. Botten, and M. J. Steel, "Confinement losses in microstructured optical fibers," *Optics Letters*, vol. 26, pp. 1660–1662, Nov. 2001.

[3.19] D. Ferrarini, L. Vincetti, M. Zoboli, A. Cucinotta, and S. Selleri, "Leakage properties of photonic crystal fibers," *Optics Express*, vol. 10, pp. 1314–1319, Nov. 2002. Available at: http://www.opticsexpress.org/abstract.cfm?URI=OPEX-10-23-1314

[3.20] D. Ferrarini, L. Vincetti, M. Zoboli, A. Cucinotta, F. Poli, and S. Selleri, "Leakage losses in photonic crystal fibers," in *Proc. Optical Fiber Communications Conference OFC 2003*, Atlanta, Georgia, USA, Mar. 23–28, 2003, paper FI5.

[3.21] J. J. Refi, "Mixing TrueWave™ RS Fiber with other Single-Mode Fibers in a Network," Bell Laboratories Innovations, Lucent Technologies, Tech. Rep., 2001.

[3.22] *Corning® SMF-28™ CPC6 Single-Mode Optical Fibre – Product Information*, Corning, 1998.

[3.23] *Ritekom G-655 Fiber – Ritekom Photonics Product Guide, 1-2*, Ritekom Photonics Corporation, 2003.

[3.24] J. T. Lizier and G. E. Town, "Splice losses in holey fibers," *IEEE Photonics Technology Letters*, vol. 13, pp. 794–796, Aug. 2001.

[3.25] K. G. Hougaard, A. Bjarklev, E. Knudsen, S. B. Libori, J. Riishede, P. M. W. Skovgaard, and J. Broeng, "Coupling to photonic crystal fibers," in *Proc. Optical Fiber Communications Conference OFC 2002*, Anaheim, California, USA, Mar. 17 –22, 2002, paper ThGG11.

[3.26] G. E. Town and J. T. Lizier, "Tapered holey fibers for spot-size and numerical-aperture conversion," *Optics Letters*, vol. 26, pp. 1042–1044, July 2001.

[3.27] B. Kuhlmey, G. Renversez, and D. Maystre, "Chromatic dispersion and losses of microstructured optical fibers," *Applied Optics*, vol. 42, pp. 634–639, Feb. 2003.

[3.28] B. T. Kuhlmey, R. C. McPhedran, C. M. de Sterke, P. A. Robinson, G. Renversez, and D. Maystre, "Microstructured optical fibers: where's the edge?," *Optics Express*, vol. 10, pp. 1285–1290, Nov. 2002. Available at: http://www.opticsexpress.org/abstract.cfm?URI=OPEX-10-22-1285

[3.29] N. A. Mortensen, J. R. Folkenberg, M. D. Nielsen, and K. P. Hansen, "Modal cutoff and the V parameter in photonic crystal fibers," *Optics Letters*, vol. 28, pp. 1879–1881, Oct. 2003.

[3.30] W. Belardi, J. H. Lee, K. Furusawa, A. Yusoff, P. Petropoulos, M. Ibsen, T. M. Monro, and D. J. Richardson, "A 10 $Gbit/s$ tunable wavelength converter based on four-wave-mixing in higly nonlinear holey fiber," in *Proc. European Conference on Optical Communication ECOC 2002*, Copenhagen, Denmark, Sept. 8–12, 2002, paper postdeadline 1.2.

[3.31] R. Tang, J. Lasri, P. Devgan, J. E. Sharping, and P. Kumar, "Microstructure-fibre-based optical parametric amplifier with gain slope of $\simeq$200 $dB/W/km$ in the telecom range," *Electronics Letters*, vol. 39, pp. 195–196, Jan. 2003.

[3.32] K. Saitoh, M. Koshiba, T. Hasegawa, and E. Sasaoka, "Chromatic dispersion control in photonic crystal fibers: application to ultra-flattened dispersion," *Optics Express*, vol. 11, pp. 843–852, Apr. 2003.

Available at: http://www.opticsexpress.org/abstract.cfm?URI=OPEX-11-8-843

[3.33] K. P. Hansen, "Dispersion flattened hybrid-core nonlinear photonic crystal fiber," *Optics Express*, vol. 11, pp. 1503–1509, June 2003. Available at: http://www.opticsexpress.org/abstract.cfm?URI=OPEX-11-13-1503

[3.34] Nonlinear photonic crystal fibers – Crystal Fibre A/S. Available at: http://www.crystal-fibre.com/products/nonlinear.shtm

# Chapter 4

# Nonlinear properties

In this chapter the nonlinear properties of PCFs are deeply analyzed. Firstly, supercontinuum generation, one of the most important applications of the fibers with enhanced nonlinear properties, is described, starting from the numerous results, both experimental and theoretical, which have been presented in literature so far.

Then, the attention is mainly directed to one particular nonlinear effect, that is the four-wave mixing, which is exploited for the optical parametric amplification. Many nonlinear triangular PCFs with different dispersion properties have been considered in order to optimize the amplifier gain performances [4.1, 4.2].

Finally, a different kind of PCF, that is the hollow-core one, has been considered, even if these fibers present negligible nonlinear characteristics. The nonlinear coefficient of hollow-core PCFs with modified honeycomb lattice has been evaluated, showing that also the nonlinear contribution of air should be taken into account [4.3].

A detailed analysis of another nonlinear effect, that is the Raman one, and its exploitation in PCFs, will be discussed apart in the Chapter 5.

## 4.1 Supercontinuum generation

Supercontinuum (SC) generation is a complex physical phenomenon which causes a significant spectral broadening of laser pulses propagating in a nonlinear medium. The SC formation through the interaction of intense pulses with matter has been discovered in the 1970s, first in condensed matter [4.4], then in single-mode fibers. In fact, by using an optical fiber as the nonlinear medium, which offers a longer interaction length and a higher effective nonlinearity, it is possible to significantly reduce the peak power, which was higher than 10 MW for bulk materials [4.5]. After the first demonstrations, the possible improvements of the SC characteristics, as well as a simplification of the technical requirements for its generation have been theoretically and experimentally investigated [4.5].

Due to its coherently pulsed nature and its high spatial brightness, SC generated in optical fibers is an ideal source for a lot of applications, like frequency metrology, $fs$-pulse phase stabilization, optical coherence tomography (OCT), ultrashort pulse compression, spectroscopy of materials and photonic structures, and fiber characterization [4.6].

### 4.1.1 Physics of supercontinuum generation

The most important aspects for the SC generation are the pulse length, the peak power, and the dispersion of the nonlinear medium with respect to the pumping wavelength, since the dispersion properties strongly influence the plethora of nonlinear effects which lead to the pulse broadening [4.7]. In particular, the positive or negative dispersion determines which kind of nonlinear effects participate in the SC formation, as well as the main characteristics of the spectrum, that is its shape and stability [4.7]. For example, in optical fibers it is necessary to choose a pulse wavelength near the zero-dispersion wavelength, so the SC is restricted in the range around 1300 nm if conventional single-mode fibers are considered. The use of dispersion-flattened or dispersion-decreasing fibers can lead to a shift of the SC spectrum towards longer wavelengths in S, C, and L band. In order to obtain SC also in the visible wavelength range, tapered fibers and, in particular, PCFs can be employed, due to their unusual dispersion properties and their enhanced effective nonlinearity [4.5]. SC spectra in PCFs have been generated with pulse widths in the range between the ten-$fs$ and the $ns$ regime, and with pump wavelengths between 532 and 1500 nm [4.8].

### 4.1.2 Highly nonlinear PCFs

A significant advance in the research regarding the SC generation has been reached with solid-core highly nonlinear PCFs. In particular, these fibers offer enhanced nonlinear properties, due to their small effective area, thus significantly reducing the peak power necessary to generate the SC. In fact, the SC formation in standard optical fibers requires pulses with an initial peak intensities more than two orders of magnitude higher with respect to the PCF case [4.9]. Moreover, by exploiting the dispersion tailoring, it is possible to properly shift the PCF zero-dispersion wavelength in the range of the Ti:Sapphire femtosecond laser systems operating around 800 nm, thus obtaining SC spectrum in the visible region [4.7]. Differently from other technologies, like amplified spontaneous emission source or incandescent lamp, SC spectra generated in PCFs offer, at the same time, a high brightness and a broad coverage [4.7]. However, it is important to underline that in conventional optical fibers, where SC is mainly generated through the self-phase modulation, the broadened spectra are symmetrical, bell-shaped-like, centered around the pump wavelength and smoother with respect to the ones obtained in PCFs [4.10]. In fact, the SC spectrum formed in PCFs is characterized by a complex shape, since a lot of different effects, such as group-velocity dispersion (GVD), self-phase modulation (SPM), cross-phase modulation (XPM), four-wave mixing (FWM), stimulated Raman scattering (SRS), birefringence, high-order soliton formation, third-order dispersion, and self-steeping, participate in the generation process [4.10]. Moreover, the large number of nonlinear processes involved in the SC generation in PCFs causes additional noise and a higher sensitivity to the amplitude fluctuations of the incident light [4.10].

Highly nonlinear PCFs are usually characterized by a high air-filling fraction and a small hole-to-hole spacing, typically in the range 1–3 μm [4.7]. These fibers can be multi-mode, with an extremely small core and a cobweb-like microstructure, like the PCF shown in Fig. 4.1, or single-mode, with a slightly larger silica core, smaller air-holes and a properly tailored zero-dispersion wavelength [4.7]. The choice of the proper PCF in order to generate the SC spectrum strongly depends on the wavelength range of the desired source and on the available pump. In particular, the zero-dispersion wavelength of the highly nonlinear PCF should be close to the center wavelength of the pump source [4.7]. Nonlinear fibers proper to femtosecond sources at 800, 1060 and 1550 nm, as well as to nanosecond at 1060 and 1550 nm can be successfully designed [4.7]. Highly nonlinear PCFs with two close zero-dispersion wavelengths have been also designed and fabricated, which open up new interesting possibilities for SC generation [4.7]. The length of the nonlinear PCF used for the SC generation is strictly related to the pump pulse length, being shorter fibers necessary for faster pulses. For example, for $fs$ pulse 1 m or less of nonlinear

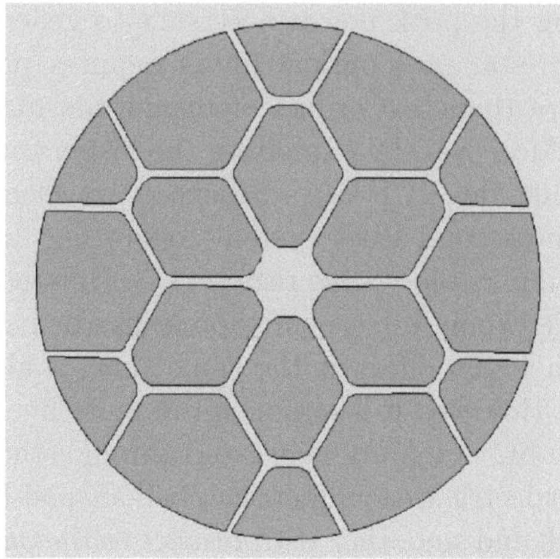

Figure **4.1**: Schematic of the cross-section of the cobweb holey fiber, proposed in [4.11]. The core diameter is 1 μm and the width of the fine silica bridges supporting the core is about 120 nm [4.12].

PCF is enough. On the contrary, a PCF 10–20 m long is necessary for the *ps* or *ns* pumping at 1060 nm [4.7].

The first demonstration of this phenomenon in a highly nonlinear PCFs has been reported in 2000 [4.13]. A solid-core triangular PCF with a core diameter of about 1.7 μm and an air-hole diameter of 1.3 μm has been employed. In particular, a 550 THz wide optical spectrum in the visible range, that is from violet to infrared, has been obtained by launching pulses of 100 *fs* duration and kW peak power around the PCF zero-dispersion wavelength, that is 770 nm [4.13].

### Highly birefringent PCFs

In order to improve the spectrum stability and to obtain the maximum broadening at a certain pump power, a polarization maintaining (PM) nonlinear PCF can be chosen. In fact, a power advantage close to a factor of two with respect to a non-birefringent fiber can be obtained by aligning the pump source polarization to one of the main axis of the PM nonlinear PCF. Moreover, the SC spectrum generated in this condition is also polarized, thus becoming useful for a wider range of applications [4.7].

The SC around 1550 nm has been demonstrated in the *Ge*-doped PM nonlinear PCF with low dispersion and low dispersion slope reported in [4.14]. Moreover, a ultrabroad SC spectrum, extending from 400 to 1750 nm, has been generated in a 5 m long highly birefringent nonlinear PCF with an effective area of 2.3 $\mu m^2$, by using a mode-locked Ti:Sapphire laser [4.15]. It has been shown that one more freedom degree in tailoring the SC characteristics is offered by the different dispersion properties of the two polarizations of the guided mode [4.15]. This consideration has been confirmed also by the experimental results presented in [4.16], which demonstrate that the SC spectrum generated at the output of a PM nonlinear PCF consists of a superposition of the spectra formed independently by the two polarizations of the guided mode.

### 4.1.3 Dispersion properties and pump wavelength

As already stated, many nonlinear processes are involved in the SC generation, causing the formation of new spectral components and their spectral broadening [4.17]. In fact, the origin of the SC generation is related to a refractive index change, due to the electric field intensity and described by the nonlinear refractive index $n_2$ [4.9]. As a consequence, a time-dependent phase is induced, which causes the generation of new spectral components at a certain spectral width around the pulse input wavelength. The efficiency of the nonlinear processes is strongly influenced by the fiber dispersion, which is responsible for the phase mismatch of different frequency components, and leads to effects like group-delay and pulse-spreading [4.17]. In particular, it is important the position of the fiber zero-dispersion wavelength $\lambda_0$ with respect to the pump wavelength $\lambda_{\text{pump}}$. For example, it has been reported that the broadest SC spectra can be obtained when $\lambda_{\text{pump}} > \lambda_0$, where the PCF dispersion is positive, that is in the anomalous dispersion regime [4.6]. On the contrary, the spectra generated for $\lambda_{\text{pump}} = \lambda_0$ or even $\lambda_{\text{pump}} < \lambda_0$ are quite narrow, but they have better flatness properties [4.6].

The property of dispersion tailoring offered by PCFs opens up a lot of new interesting possibilities to exploit for SC generation.

#### Pump in the anomalous dispersion region

Since in highly nonlinear PCFs the zero-dispersion wavelength can be shifted to the visible region, the typical wavelength of a fs Ti:Sapphire laser system,

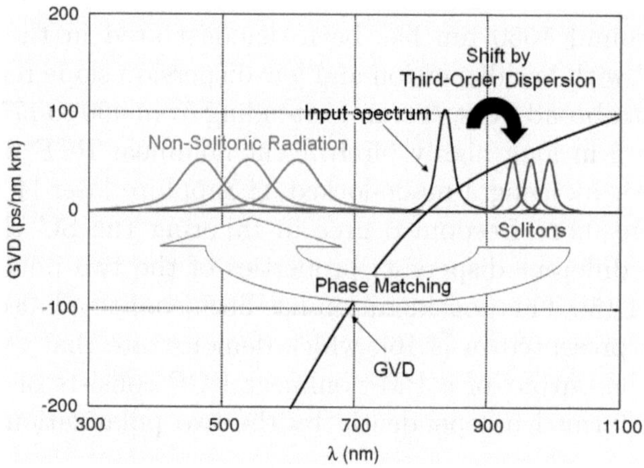


Figure **4.2**: Scheme of the SC formation by fission of higher-order solitons, as described in [4.9].

that is around 800 nm, falls in the fiber anomalous dispersion region, where the pump pulse turns into a higher-order soliton. In fact, in the anomalous dispersion region the balance between the GVD and the SPM is responsible for the formation of solitons, whose order $N$ increases with the pulse amplitude. These higher-order solitons are not stable, due to the effects of third-order dispersion, which are higher in PCFs with respect to conventional optical fibers [4.9], intrapulse Raman scattering and self-steepening, so they break up in their constituent first-order solitons, as shown in Fig. 4.2. During the decay, in order to maintain its shape, every soliton emits a blue-shifted nonsolitonic radiation at a wavelength which depends on the phase-matching condition with the pulse itself, and at the same time it shifts to the infrared range, until reaching the stability [4.8, 4.9]. In this way, a gap in the spectrum is formed around the zero-dispersion wavelength [4.18]. After this mechanism, which provides the initial spectral broadening, a complex interaction among FWM, SRS, and dispersion of the fiber causes the formation of a broad and flat SC [4.8,4.16]. These additional nonlinear processes have the positive effect to help the spectrum flattening, filling the gap between the solitons, and the nonsolitonic radiation in the visible spectral range [4.16].

It is important to underline that the anomalous dispersion, which is necessary for the SC generation by soliton fission, is also responsible of the high susceptibility of the broadened spectrum to the input pulse noise, which is amplified by modulation instabilities [4.19]. Other parameters of the input pulse can significantly influence the properties of the SC generated [4.19]. In fact, it has been demonstrated that the power and the chirp of the pulses, as well as the linear properties of the PCF, that is its modal index and GVD, significantly affect this SC generation mechanism and the spectrum characteristics [4.20]. The SC spectra reported in literature usually do not extend to the wavelengths lower than 380 nm [4.21]. The relative position of the pump wavelength with respect to the zero-dispersion wavelength represents the main limiting factor for the spectral broadening in the SC generation [4.21]. Broader spectra can be generated by shifting the pump away from the zero-dispersion wavelength, at the expense of the gap widening and of a drastic reduction of the blue-wavelength components [4.18, 4.21]. A possible alternative is to increase the pump power, which leads to the merging of the different spectral components [4.18].

### Pump in the normal dispersion region

The first results regarding the SC generation in a highly nonlinear PCF with pumping in the normal dispersion regime have been presented in 2001 [4.22]. A smooth and stable SC spectrum has been obtained, which is suitable for pulse compression and OCT [4.7, 4.22].

Differently from the case previously described, when the pump is in the normal dispersion regime and fs pulse are considered, the SPM becomes the most important nonlinear effect for the SC generation, while the spectral broadening towards the longer wavelengths is provided by the SRS [4.7, 4.8]. The shape and the bandwidth of the SC spectrum generated in this condition are strongly influenced by the pump position with respect to the zero-dispersion wavelength and by the pump power [4.7, 4.18]. In particular, as the pump is shifted closer to the zero-dispersion wavelength, other nonlinear processes, like FWM, contribute to the SC generation and broader spectra can be formed, as shown in [4.7] for a 2.5 μm core PCF with zero-dispersion wavelength around 900 nm. In fact, even for the pumping in the normal dispersion region, when the pump power increases, an asymmetry is introduced in the spectrum, due to the high dispersion slope and the SRS, and solitons are formed as soon as the spectrum is broadened beyond the zero-dispersion wavelength, that is in the anomalous dispersion regime [4.7]. For example, when the pump wavelength is fixed at 800 nm, a soliton is generated around 940 nm, whose self-frequency is shifted to the longer wavelengths for increasing pump power values [4.7].

For longer pump wavelengths, closer to the zero-dispersion wavelength, multiple solitons are formed, whose positions depend on the pump power, and a higher instability is introduced into the spectrum, which is more sensible to the changes of the coupling efficiency and the power fluctuations in the pump laser [4.23]. The influence of the pump power on the SC formation is shown in [4.23] for the same PCF and a fixed pump at 875 nm. For the higher pump power values new spectral components have been observed also at the lower wavelengths with respect to the pump wavelength, which are probably due to the FWM processes involving the solitons [4.23].

### Pump between two zero-dispersion wavelengths

As already described, it is possible to tailor the PCF dispersion properties by properly changing the geometry of the fiber cross-section. In particular, it is possible to design highly nonlinear PCFs with two close zero-dispersion wavelengths, which can be successfully exploited for SC generation. By choosing a pump wavelength between the two zero-dispersion wavelengths, stable, and compressible spectra with a high spectral density and low noise can be generated [4.19]. As demonstrated by experimental measurements, the SC spectra generated in PCFs with these dispersion properties are characterized by two peaks at each side in the normal dispersion region [4.7, 4.19].

Differently to what happens in PCFs with only one zero-dispersion wavelength or with two widely separated dispersion wavelengths, in these nonlinear fibers SPM is responsible for the initial spectral broadening, thus providing the seed for the FWM process, both degenerate and nondegenerate [4.7, 4.19]. As soon as the intensity is low enough to satisfy the phase-matching condition, FWM becomes effective for the SC generation [4.7,4.19]. Since the soliton dynamics have a minor role in the SC formation, a lower noise is contained in the generated spectrum.

The SC obtained with PCFs with two zero-dispersion wavelengths presents only a slight dependence on the input pulse on a wide range of pulse characteristics [4.19]. By tuning the pump in the range between the zero-dispersion wavelengths, the light is generated in the same two wavelength ranges, even if the ratio between the two peaks can vary [4.7]. Moreover, since the center wavelength of the two peak is related to the zero-dispersion wavelengths, a desired SC spectrum for a certain application can be obtained by properly designing the PCF dispersion properties [4.7]. For example, the nonlinear PCF proposed in [4.19] has the zero-dispersion wavelengths at 780 and 945 nm.

As demonstrated in [4.19], the sharp inner edges of the two peaks of the SC spectrum generated with 40 fs pulses at 790 nm are fixed at about 740 and 950 nm, while the outer edges slowly moves outwards when the pulse energy increases [4.19]. It is important to underline that, whatever the input pulse characteristics, that is the wavelength, the energy and the chirp, the pump is very effective, being almost complete the depletion of the power between 740 and 950 nm, which is contained into the two peaks [4.7, 4.19].

The physical mechanisms underlying the SC generation in this kind of nonlinear PCFs explained so far apply for the fiber reported in [4.19], which presents a separation of about 165 nm between the two dispersion wavelengths. On the contrary, it has been demonstrated that in a PCF with a wider spacing between the two zero-dispersion wavelengths, that is around 700 nm, the most important processes for the SC formation are the amplification of dispersive waves and the soliton self-frequency shift (SSFS) [4.21]. In fact, due to the SSFS, the center wavelengths of the multiple solitons split from the initial pump pulses, which are close to the zero-dispersion wavelength in the visible, shift towards the zero-dispersion wavelength in the infrared, that is $\lambda_{ZDI}$ in Fig. 4.3, thus acting as pumps for the amplification of dispersive waves beyond

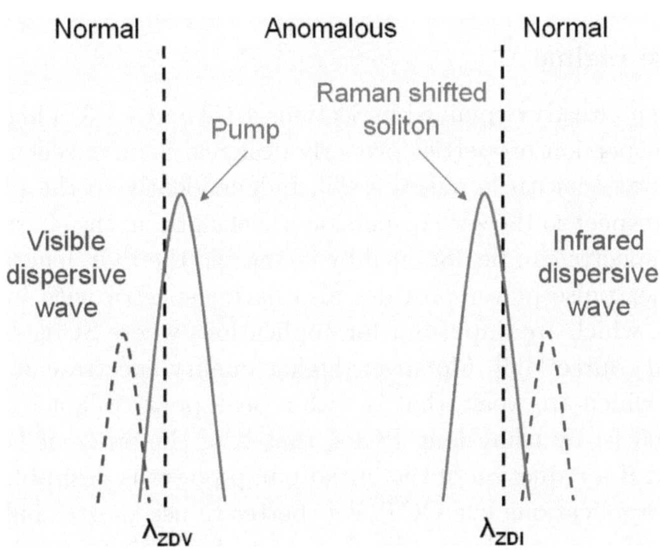


Figure **4.3**: Scheme of the dispersive wave amplification in a two zero-dispersion wavelength PCF, as described in [4.21]. $\lambda_{ZDI}$ and $\lambda_{ZDV}$ are the zero-dispersion wavelengths in the infrared and the visible, respectively.

$\lambda_{ZDI}$ [4.21]. This mechanism of SC generation, described in [4.21] for a nonlinear PCF with zero-dispersion wavelength at 690 and 1390 nm, represents an interesting possibility to increase the SC bandwidth in the infrared, without negatively affecting the blue wavelength components [4.21]. Obviously, the optimum position of the second zero-dispersion wavelength, that is the one in the infrared, is influenced by the necessity of the soliton spectrum to extend towards $\lambda_{ZDI}$. In particular, the position of $\lambda_{ZDI}$ with respect to the pump wavelength is related to the input pulse energy, the fiber nonlinear properties and its confinement losses in the infrared wavelength range [4.21].

### 4.1.4 Influence of the pump pulse regime

In the first experiments of SC generation in PCFs high power fs pulses have been used. However, it has been demonstrated that the spectral broadening in this kind of fibers can be obtained also with ps and ns pulses. In these conditions SPM becomes a negligible effect, while SRS and parametric FWM participate in the SC formation [4.24]. Telecommunication sources more cost-effective with respect to the very expensive fs laser systems, such as fiber amplifiers, can be employed for the SC generation in nonlinear PCFs [4.7].

#### Short pulse regime

As it has been already explained in Sections 4.1.2 and 4.1.3, a highly nonlinear PCF with dispersion properties properly designed is an excellent medium for the SC generation with fs pulses [4.25]. Independently to the position of the pump with respect to the zero-dispersion wavelength, in the short pulse regime broader SC spectra can be obtained by increasing the PCF length or the pulse power. Higher pulse power provides also flatter spectra with lower intensity fluctuations, which are important for applications where SC is simply used as a broadband source [4.6]. Moreover, higher quality spectra can be generated with pulses which are weak, that is with a peak power of some kW, and long that is of 100 fs, by using long PCFs, that is of the order of 1 m [4.10]. On the contrary, if a reduction of the initial pump power is desirable, as for some commercial applications like OCT, it is better to use shorter pulses [4.10].

In most of the experiments performed so far with fs pulses Ti:Sapphire laser systems operating around 800 nm with pulse energies of several nJ or more have been used. Recently, also erbium-doped fiber lasers around 1560 nm have been considered as compact diode-pumped sources for the SC formation, since they provide some advantages. In fact, it is possible to transfer the SC generation technology towards the telecommunication window, that is the C band centered at 1550 nm, with these fiber lasers, which are also more compact and more reliable with respect to bulky Ti:Sapphire laser systems [4.25]. A drawback of this kind of sources is the fact that it is necessary to amplify the pulse energy to the high levels adequate for the SC generation [4.25].

Up to now, an erbium-doped fiber laser has been used to generate a broad SC spectrum only in a silica highly nonlinear PCF with zero-dispersion wavelength around 1500 nm. However, with this PCF high energy pulses or fiber lengths of several meters are necessary for the SC generation. As an alternative, an extruded PCF made with SF6 glass, which is characterized by a higher nonlinear refractive index and different dispersion properties with respect to the silica fiber, has been used to generate an octave-broad SC spectrum [4.25].

### Long pulse regime

The SC formation in PCFs is possible also by considering longer pulses. In this case the SC generation is the result of the formation of new spectral components through the SRS or the FWM, with a subsequent broadening due to the merging of these spectral components, while the SPM of the pump pulses is negligible [4.26]. This mechanism for the SC formation requires less expensive sources, but the generated spectrum is affected by the asymmetry typical of the SRS and by the inefficiency related to the phase-matching parametric processes [4.26].

The cobweb PCF reported in Fig. 4.1 has been used to generate a broad SC spectrum with long pump pulses, that is broader than 10 ps, of sub-kilowatt power at 647 nm, that is in the normal dispersion regime, as reported in [4.26]. It is important to underline that, for the maximum peak power, the SC extends from 400 to beyond 1000 nm, covering all the visible spectral range and the near infrared, so the use of intense ultrashort pulses is not necessary for an efficient SC formation in highly nonlinear PCFs [4.26]. However, the position of the pump wavelength with respect to the zero-dispersion wavelength is very important for the SC formation when long pump pulses are employed [4.26].

Long pulses, that is in the ns regime, around 1060 nm can be used for the spectral broadening, which is dominated by SRS [4.7]. Since the generation of spectral components at wavelengths shorter than 1000 nm is difficult to obtain, the pump at 1064 nm can be combined with its second harmonic at 532 nm. In this way a SC spectrum two-octave wide has been formed in a nonlinear PCF by using ns pulses with a peak power of about 3 kW [4.7]. In fact, besides SRS, the FWM between the two pumps is responsible for the presence of the new spectral components at the wavelengths lower than 532 nm [4.7].

PCFs can be designed with two-zero dispersion wavelengths also in the telecommunication wavelength range. For example, the nonlinear fiber reported in [4.27] is characterized by two zero-dispersion wavelengths around 1475 and 1650 nm [4.7]. In order to obtain the SC spectrum, the nonlinear PCF has been pumped at a wavelength between the two zero-dispersion wavelengths, that is at 1555 nm, with 2 ps pulses. Looking at the spectra reported in [4.7] for different pump power values, it is possible to notice significant differences with respect to the results obtained in PCFs with similar dispersion properties pumped by fs pulses around 800 nm, shown, for example, in [4.21]. In particular, even if the two peaks are still present in the generated spectrum, the pump power in the range between the two zero-dispersion wavelength is not depleted, due to the lower values of the input pulse power and of the PCF nonlinear coefficient in the C band [4.7]. In these conditions the SPM process is significantly reduced and, consequently, also the FWM efficiency decreases. As a result, the spectrum is characterized by a large peak, due to the residual pump light, and by spectral components at long wavelengths, due to the SRS. With shorter pulses the SPM process is more effective and reduce the pump power, even if an almost complete pump depletion is possible only at shorter wavelengths, that is in the 800 nm region [4.7].

### 4.1.5 Applications

The most important application of the SC spectra is the replacement of the white light sources, which are usually tungsten-based, in different characterization setups, such as for spectroscopy, microscopy, interferometer-based dispersion measurements, and broadband attenuation measurements [4.7]. SC sources, which are characterized by the spectral width of a tungsten lamp and the intensity of a laser, solve the problems of the traditional incandescent sources, that is the low brightness and the coupling inefficiency to optical fibers [4.7]. Moreover, these new sources can drastically improve the signal-to-noise ratio, reduce the measurement time or widen the spectral range where the measurements can be made [4.28]. Most of the SC sources experimentally realized have an output power in the mW range, but also higher output power values have been obtained [4.7].

The main negative aspect of the SC sources with respect to the traditional incandescent ones is the high cost of their pump, which can be, in the extreme case, a very expensive large fs system [4.7]. Consequently, more compact and cost-effective schemes for the SC generation should be developed, by taking into account, for example, the sources around 1060 nm with long pulses in the ns and ps regime [4.7].

One of the most important applications of the SC sources is the OCT, a new technology, based on low-coherence interferometry, used for *in vivo* and *in situ* cross-sectional morphological imaging of transparent and non-transparent biological tissue on a micrometer scale [4.29]. OCT requires smooth spectra, that is variations of less than 10 dB, since spectral gap can affect the image quality and the measurement precision [4.10]. While broad spectra which extend also into the visible range down to 400 nm are necessary to provide access to wavelengths interesting for spectroscopic OCT of biological chromophores [4.10], the spectral region between 1200 and 1500 nm is particularly important for the OCT, since it permits high penetration depth in biological tissues and spectrally resolved imaging of the water absorption bands [4.29]. However, the OCT longitudinal resolution is inversely proportional to the source width and proportional to the square of the central wavelength, so it becomes poor at long wavelength for sources which are not wide enough [4.7]. In standard OCT systems sources based on the amplified spontaneous emission from doped fibers or semiconductors, or superluminescent diodes are considered as light sources, usually providing a longitudinal resolution of 10–15 μm [4.7, 4.29]. All these sources suffer limited bandwidth and restricted wavelength range [4.7]. On the contrary, as it has been already demonstrated, the SC spectra generated in PCFs are characterized by enormous bandwidths, thus providing an unprecedented resolution [4.7]. In fact, it has been demonstrated the use of the SC generated in a PCF in a OCT system for *in vivo* imaging of biological tissue, reaching for the first time a resolution of 2.5 μm in the wavelength range around 1300 nm [4.29]. The SC sources obtained in PCFs with slow pulses around 1060 nm are particularly promising for the OCT, because the large flat spectrum mainly generated by SRS is very stable and can be filtered in order to select the desired wavelength range [4.7].

The octave-spanning frequency comb which can be generated in PCFs with fs pulses has provided significant advantages in frequency metrology [4.9]. In particular, frequency standards based on SC have been one of its first applications to be commercialized [4.7].

SC sources offer important improvements also for the low-coherence white-light interferometry, useful for displacement measurements, for the position determination of flaws in optical waveguides, and for the chromatic dispersion measurement in optical fibers and planar waveguides, because the spatial resolution obtainable increases with the source bandwidth [4.28].

Due to their high degree of spatial coherence, SC sources are useful in spectroscopy for the fast data acquisition on small-volume samples in biology, chemistry, medicine, physics, or environmental monitoring. In fact, the SC light can be focused into a small spot, or collimated in a narrow beam for long-path length measurements in analytes with low absorbance [4.28].

## 4.2 Optical parametric amplification

Parametric amplification provides a new possibility to amplify signals in optical transmission systems, besides erbium-doped or Raman fiber amplifiers. The parametric gain is based on highly efficient FWM, relying on the relative phase between four interacting photons [4.30–4.32]. By pumping the fiber with an intense wave, a wide and flat gain spectrum can be obtained over two bands surrounding the pump wavelength. Modern high-power sources have increased the interest in optical parametric amplifiers (OPAs), whose gain bandwidth can be tailored to operate at any wavelength, providing amplification outside the conventional erbium-doped one. Besides broadband amplification at arbitrary wavelength, the parametric process offers a variety of applications, such as, for example, wavelength conversion, pulse reshaping and soliton–soliton interaction [4.30]. Multiple pump schemes can further enhance the OPA efficiency, both in terms of maximum gain and bandwidth [4.33].

Fiber nonlinearity and dispersion are fundamental aspects for a successful OPA design. In fact, to achieve high and broadband gain in OPAs, the phase-matching condition demands a low dispersion slope, while the efficiency of the nonlinear process requires a small fiber effective area, in order to have a high nonlinearity. In the last few years, highly nonlinear optical fibers with nonlinear parameter five to ten times higher than that of conventional fibers have been introduced, and OPA gains up to 50 dB have been experimentally demonstrated [4.34].

PCFs are very interesting for optical parametric amplification [4.35], since they can significantly enhance the FWM process [4.36–4.39], whose conversion efficiency is strictly related to the dispersion and the nonlinear properties of the optical fiber through the phase-matching condition [4.30]. In fact, PCFs offer the possibility to engineer the zero-dispersion wavelength, the dispersion curve and the nonlinear coefficient value. In particular, highly nonlinear PCFs [4.40], as well as fibers with very flat dispersion curves around the zero-dispersion wavelength, have been successfully designed [4.41]. Also PCFs with both these properties have been proposed [4.42].

In the study reported here the two kinds of all-silica triangular PCFs with flattened dispersion curve, zero-dispersion wavelength around 1550 nm and high nonlinear coefficient described in Chapter 3 have been considered. In order to show that triangular PCFs have interesting properties for parametric amplification, the phase-matching condition has been analyzed by varying the parameters which define the fiber cross-section geometry. The present analysis has been performed by means of the full-vector modal solver based on the FEM, described in Appendix A. Simulation results have demonstrated that the possibility offered by triangular PCFs to engineer the dispersion curve, as well as the nonlinear coefficient value, can be successfully exploited to satisfy the FWM phase-matching condition, with positive consequences on the gain bandwidth of the parametric amplification process [4.1, 4.2].

### 4.2.1 Triangular PCFs for OPA

The first fiber type, described in Section 3.3.1, has been designed by changing the diameter $d_1$, $d_2$, and $d_3$ of all the air-holes belonging to the first three rings around the core, as it is shown in Fig. 3.14. It has been already demonstrated that a proper choice of $\Lambda$, $d$, $d_1$, $d_2$, and $d_3$ can provide very high nonlinear coefficient values and very flat dispersion curves. In particular, the following geometric parameters have been chosen, that is $\Lambda = 0.9$ μm, $d = 0.81$ μm, $d_1 = 0.42\Lambda$, $d_2 = 0.87\Lambda$, and $d_3 = 0.86\Lambda$. The dispersion curve of this PCF with three modified air-hole rings is shown in Fig. 3.19a.

In the second triangular PCF type here considered, shown in Fig. 3.21, three of the air-holes belonging to the first ring have a different diameter $d_f < d$. As reported in Section 3.3.2, due to the shape of the silica core, these fibers are called triangular-core PCFs. Among the PCFs previously designed with the best dispersion and nonlinear properties, that is the ones with the dispersion curves reported in Fig. 3.25, three fibers have been considered in the present analysis, as reported in Table 4.1.

Table 4.1: Zero-dispersion wavelength and dispersion slope of the designed PCFs of the first (first row) and the second (last three rows) type [4.1, 4.2].

| $\Lambda$ (μm) | $d$ (μm) | $d_f$ (μm) | $\lambda_0$ (nm) | $S_0$ (ps/km · nm$^2$) |
|---|---|---|---|---|
| 0.9 | 0.81 | – | 1510.5 | $-0.94 \cdot 10^{-2}$ |
| 1.4 | 0.55 | 0.0 | 1540.2 | $-3.87 \cdot 10^{-2}$ |
| 1.6 | 0.65 | 0.29 | 1550.5 | $-1.8 \cdot 10^{-2}$ |
| 1.7 | 0.65 | 0.32 | 1563.3 | $-1.3 \cdot 10^{-2}$ |

Table 4.2: Nonlinear coefficient $\gamma$ for the PCFs of the first (first row) and the second (last three rows) type [4.1, 4.2].

| $\Lambda$ (μm) | $d$ (μm) | $d_f$ (μm) | $\gamma$ (W · km)$^{-1}$ |
|---|---|---|---|
| 0.9 | 0.81 | – | 40.73 |
| 1.4 | 0.55 | 0.0 | 10.92 |
| 1.6 | 0.65 | 0.29 | 10.97 |
| 1.7 | 0.65 | 0.32 | 9.7 |

### Dispersion and nonlinear properties

The zero-dispersion wavelength $\lambda_0$ and the dispersion slope $S_0$ values of the fibers analyzed in the present study are summarized in Table 4.1. The values for the fiber of the first type are shown in the first row, while the ones for the triangular-core PCFs are reported in the last three rows. Notice that for the PCF with $\Lambda = 0.9$ μm the zero-dispersion wavelength occurs at a slightly lower value, that is 1510.5 nm, and the dispersion slope is very low, as well.

The nonlinear coefficient values at 1550 nm, evaluated according to Eq. (A.8) in Appendix A, are reported in Table 4.2 for the studied PCFs. Notice that the triangular-core PCF with the lowest slope has also the lowest $\gamma$ value. This suggests that for this kind of fibers a proper trade-off between the dispersion slope and the nonlinear coefficient values must be found. For this reason, considering the PCFs of the second type, the attention will be focused on the first and the second fiber in Tables 4.1 and 4.2, that is the fiber with $\Lambda = 1.4$ μm, $d = 0.55$ μm, $d_f = 0$ μm, and the one with $\Lambda = 1.6$ μm, $d = 0.65$ μm, $d_f = 0.29$ μm, which have almost the same nonlinear coefficient value. As already observed, the fiber of the first type has been selected for its high $\gamma$ value, that is 40.73 (W·km)$^{-1}$, which is due to its small pitch.

### 4.2.2 Phase-matching condition in triangular PCFs

In order to show how the triangular lattice PCFs here proposed can be successfully used for optical parametric amplification, the phase-matching condition has been analyzed. Under the assumption of undepleted pump, this condition reads as

$$\kappa = \Delta\beta + 2\gamma P_{\mathrm{p}} = 0 \ , \qquad (4.1)$$

where $\kappa$ is the phase-mismatch parameter, $\Delta\beta$ is the linear wave-vector mismatch and $P_{\mathrm{p}}$ is the pump power [4.30]. When this condition is satisfied, the maximum gain can be obtained through the parametric amplification, since the power flow from the pump at $\lambda_{\mathrm{p}}$ to the signal at $\lambda_{\mathrm{s}}$, which are involved in the FWM process, is highly efficient [4.30]. The phase-matching is obtained when the nonlinear phase shift $2\gamma P_{\mathrm{p}}$ is compensated by a negative $\Delta\beta$ [4.30]. The linear component of the phase-mismatch parameter can be calculated by expanding in Taylor series the phase constant $\beta(\omega)$ around the zero-dispersion frequency $\omega_0 = 2\pi c/\lambda_0$, that is,

$$\Delta\beta = \left\{\beta_3(\omega_{\mathrm{p}} - \omega_0) + \frac{\beta_4}{2}\left[(\omega_{\mathrm{p}} - \omega_0)^2 + \frac{1}{6}(\omega_{\mathrm{p}} - \omega_{\mathrm{s}})^2\right]\right\}(\omega_{\mathrm{p}} - \omega_{\mathrm{s}})^2 \ , \qquad (4.2)$$

where $\beta_3$ and $\beta_4$ are, respectively, the third and fourth derivative of $\beta(\omega)$ calculated at $\omega_0$, $\omega_{\mathrm{p}}$ is the pump frequency and $\omega_{\mathrm{s}}$ the signal one. In the present analysis the contribution from $\beta_4$ has been considered, as shown in Eq. (4.2). In fact, when taking into account PCFs, the waveguide contribution to the dispersion curve is significant, thus higher-order derivatives of $\beta(\omega)$ are usually larger than in conventional fibers, and they can not be neglected [4.37, 4.38].

The values of $\beta_3$ and $\beta_4$ for the three fibers considered in the present study are reported in Table 4.3. Notice that the first type PCF, that is the one with $\Lambda = 0.9$ μm, has $\beta_3$ and $\beta_4$ which are, respectively, about five and two times lower than those of the triangular-core PCFs with $\Lambda = 1.4$ μm and

Table 4.3: Dispersion properties of the PCFs of the first (first row) and the second (last two rows) type [4.1, 4.2].

| **$\Lambda$ (μm)** | $d$ (μm) | $d_{\mathrm{f}}$ (μm) | $\beta_3$ **(ps$^3$/km)** | $\beta_4$ **(ps$^4$/km)** |
|---|---|---|---|---|
| 0.9 | 0.81 | – | $-1.38 \cdot 10^{-2}$ | $1.40 \cdot 10^{-4}$ |
| 1.4 | 0.55 | 0.0 | $-6.31 \cdot 10^{-2}$ | $5.57 \cdot 10^{-4}$ |
| 1.6 | 0.65 | 0.29 | $-2.84 \cdot 10^{-2}$ | $2.99 \cdot 10^{-4}$ |

$\Lambda = 1.6$ µm. It is important to underline that the values of $\beta_3$ and $\beta_4$ have been evaluated by deriving the $8th$ order polynomial fitted to the dispersion curve. Moreover, their accuracy has been checked following a second approach, besides the expression given by Eq. (4.2). In particular, the linear wave-vector mismatch has been calculated also through the relation

$$\Delta\beta = \beta(\omega_s) + \beta(\omega_i) - 2\beta(\omega_p) \; , \tag{4.3}$$

being $\beta(\omega_s)$, $\beta(\omega_i)$, and $\beta(\omega_p)$ the phase constant, respectively, of signal, idler and pump, which have been obtained by the FEM solver, as described in Appendix A. The agreement between the two approaches is very good.

The linear wave-vector mismatch versus the wavelength difference between the signal and the pump, $|\lambda_s - \lambda_p|$, has been calculated in the range 0–60 nm for the three considered PCFs. For example, Fig. 4.4a and b report two sets of $\Delta\beta$ curves obtained by choosing different $\lambda_p$, in order to get similar values of the $\Delta\beta$ minimum, that is $-5$ and $-10\,\mathrm{km}^{-1}$, respectively, and thus to compare the properties of the three different PCFs. Notice that, being $\beta_3$ negative for all the considered PCFs, $\omega_p$ must be greater than $\omega_0$, that is $\lambda_p < \lambda_0$, in order to obtain a negative $\Delta\beta$, as shown in Eq. (4.2). It is important to underline the presence of two symmetrical minima in all the $\Delta\beta$ curves, which are due to the different sign of $\beta_3$ and $\beta_4$. In fact, according to Eq. (4.2), as $|\lambda_s - \lambda_p|$ increases, the positive contribution of $\beta_4$ to the linear wave-vector mismatch becomes higher, until it dominates the negative one provided by $\beta_3$. As a consequence, $\Delta\beta$ decreases initially when $\lambda_s \simeq \lambda_p$, it reaches a negative minimum value and then it increases, becoming positive and no longer useful for satisfying the phase-matching condition. Moreover, results have shown that the value and the position of the minimum are strictly related to the pump wavelength. In particular, it becomes more negative and further from $\lambda_p$ when $|\lambda_p - \lambda_0|$ increases, as it can be easily observed for all the considered PCFs by comparing Fig. 4.4a with b.

Looking at the three $\Delta\beta$ curves shown in Fig. 4.4a or b, it is possible to notice the influence of $\beta_3$ and $\beta_4$ on the linear wave-vector mismatch. In particular, the minima of the linear wave-vector mismatch, as well as the condition $\Delta\beta = 0$, can be obtained for greater values of $|\lambda_s - \lambda_p|$ if the PCF with the lowest $\beta_3$ and $\beta_4$ is considered, that is the first type one with $\Lambda =$ 0.9 µm. Moreover, the same minima values of $\Delta\beta$ can be reached by choosing a pump wavelength further from $\lambda_0$. For example, by considering the minimum equal to $-10\,\mathrm{km}^{-1}$, $\lambda_p - \lambda_0 \simeq -1.95$ nm for the first-type PCF with $\Lambda =$

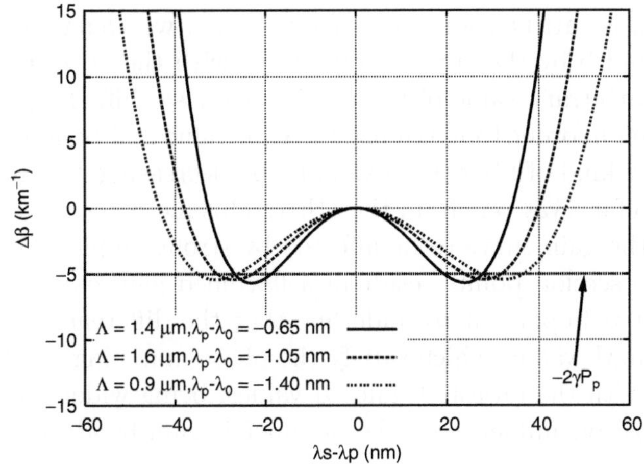


(a)

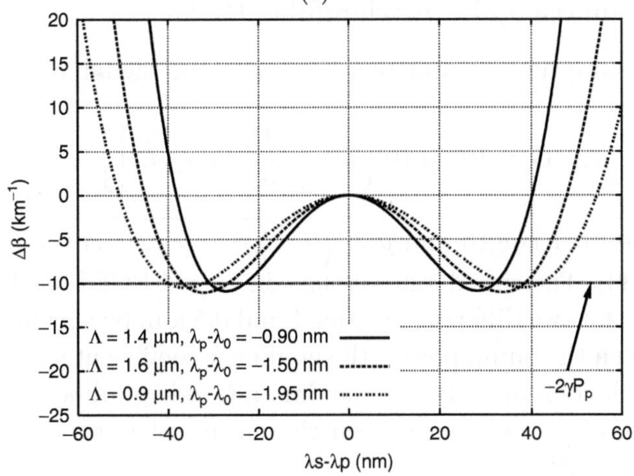


(b)

Figure **4.4**: Linear wave-vector mismatch $\Delta\beta$ versus $\lambda_s - \lambda_p$ for the three PCFs of Table 4.3. The pump wavelength $\lambda_p$ has been chosen to obtain $\Delta\beta$ minima of about (a) $-5$ km$^{-1}$ and (b) $-10$ km$^{-1}$ [4.2].

0.9 μm, $\lambda_p - \lambda_0 \simeq -1.5$ nm for the triangular-core PCF with $\Lambda = 1.6$ μm, while $\lambda_p - \lambda_0 \simeq -0.9$ nm for the one with $\Lambda = 1.4$ nm.

Finally, it is important to underline that, for $2\gamma P_p$ values lower than the absolute value of the $\Delta\beta$ minimum, the phase-matching condition of Eq. (4.1), that is $\kappa = 0$, is satisfied for two different signal wavelengths. This is shown in Fig. 4.4a and b by the curve intersections with the horizontal lines, which represent an arbitrary value of the nonlinear phase shift $2\gamma P_\mathrm{p}$. Obviously, $P_\mathrm{p}$ must be chosen in order to maximize the gain value and bandwidth, according to the selected kind of fiber. For example, by decreasing $P_\mathrm{p}$, the two intersection points go far away one from the other. This increases the bandwidth, but it can cause the gain curve to be affected by strong ripples. On the contrary, two close intersection points result in a flattened gain curve with a reduced bandwidth. It is important to underline that the difference between the two signal wavelength values which satisfy the phase-matching condition is higher for the PCF with the lowest $\beta_3$ and $\beta_4$ values, being wider its $\Delta\beta$ minimum. This has a positive influence on the parametric gain bandwidth.

### Optical parametric gain in triangular PCFs

Under the assumption of undepleted pump, the signal power gain can be expressed as

$$G = 10 \log 10 \left\{ 1 + \left[ \frac{\gamma P_p}{g} \sinh(gL) \right] \right\} , \tag{4.4}$$

where $L$ is the fiber length and $g = \sqrt{(\gamma P_p)^2 - (\kappa/2)^2}$ is the parametric gain coefficient [4.30]. The signal gain of the considered PCFs has been calculated versus $|\lambda_\mathrm{s} - \lambda_\mathrm{p}|$ for two different lengths, 1 and 0.5 km, by varying $P_\mathrm{p}$ so that the product between the pump power, the nonlinear coefficient and the fiber length is constant and, consequently, the maximum $G$ is kept almost fix [4.30]. In the present analysis $P_\mathrm{p}$ values have been chosen in order to provide a maximum gain of about 16 dB.

As shown in Fig. 4.5, a very flat gain can be obtained over a wide signal wavelength range with all the triangular PCFs here considered. Notice that the bandwidth is wider when the phase-matching condition is satisfied with the most negative $\Delta\beta$ value, that is $-10\,\mathrm{km}^{-1}$, as reported in Fig. 4.5b. Moreover, a larger gain bandwidth can be obtained with the PCF characterized by the lowest $\beta_3$ and $\beta_4$ for both the amplifier configurations reported in Fig. 4.5a and b. In particular, considering the fiber of the first type, a 3 dB-bandwidth of 30 and 35 nm has been reached, by satisfying the phase-matching condition, respectively, for a nonlinear phase shift of 5 $\mathrm{km}^{-1}$, corresponding to $P_\mathrm{p} =$ 0.062 W and a fiber length equal to 1 km, and a nonlinear phase shift of 10 $\mathrm{km}^{-1}$, corresponding to $P_\mathrm{p} = 0.124$ W and a fiber length equal to 0.5 km.

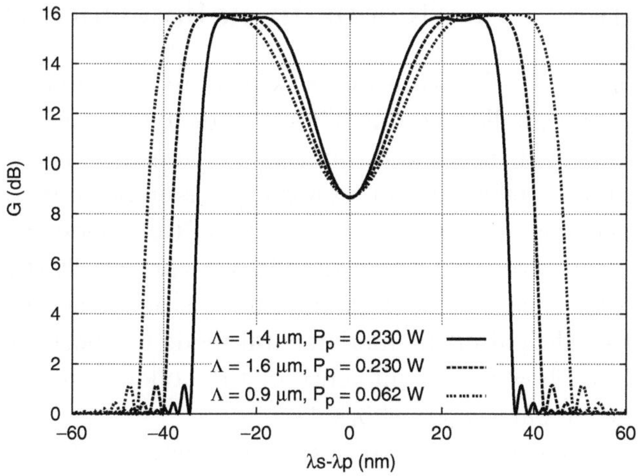


(a)

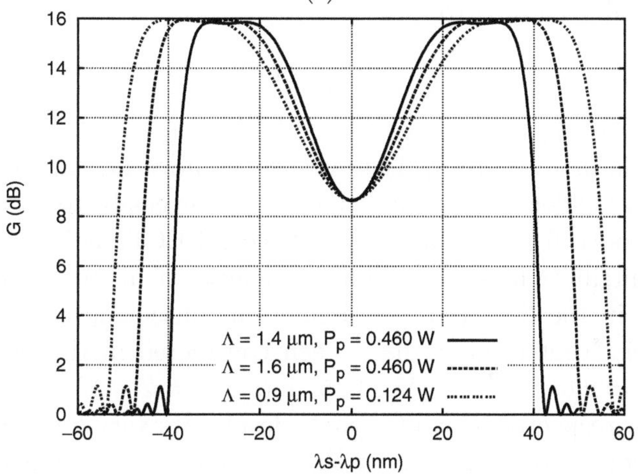


(b)

Figure **4.5**: Signal power gain $G$ versus $\lambda_s - \lambda_p$. The pump power level used for each PCF is indicated in the figure label. All the fibers are (a) 1.0 km and (b) 0.5 km long, respectively [4.2].

Notice that the $P_p$ values used for the first type PCF are about four times lower than those necessary for the two triangular-core fibers, being 0.230 and 0.460 W, respectively. This is due to the difference among the nonlinear coefficient values of the three triangular PCFs, as reported in Table 4.2.

## 4.3 Nonlinear coefficient in hollow-core PCFs

Differently from small silica-core PCFs, which are characterized by enhanced nonlinear properties, hollow-core PBGFs provide a very small overlap between the guided-mode field distribution and silica, thus allowing to dramatically reduce the nonlinear effects [4.43]. However, it is important to give an exact evaluation of the nonlinear characteristics also of this particular kind of PCFs. To this aim, the four modified honeycomb PBGFs previously studied in Section 2.3 have been considered for a thorough analysis of the nonlinear properties. The nonlinear coefficient of the fundamental mode has been calculated for all the PBGFs, thus showing the influence of the hollow core, as well as of the cladding geometric parameters [4.3, 4.44].

The nonlinear coefficient $\gamma$ describes the change of the fundamental mode phase constant $\beta$ due to the nonlinear effects for a given input power $P$. It has been calculated according to

$$\gamma = \frac{\delta\beta}{P} = \frac{2\pi}{\lambda}\frac{\delta n_{\text{eff}}}{P}, \tag{4.5}$$

being $\beta = k_0 \cdot n_{\text{eff}}$, where $k_0$ is the wave number in the vacuum and $n_{\text{eff}}$ is the effective index of the guided mode. Conversely to what happens in conventional optical fibers, in the hollow-core PBGFs it is necessary to separate the contribution to the nonlinear effects of air and silica, which are both present in the fiber cross-section. As a consequence, two values of the nonlinear-index coefficient $n_2$ have been considered in the present analysis, that is $n_{2_{\text{SiO}_2}} = 2.6 \cdot 10^{-20}$ m$^2$/W [4.45] and $n_{2_{\text{air}}} = 2.9 \cdot 10^{-23}$ m$^2$/W [4.46] for silica and air, respectively. Notice that, despite its very low nonlinearity, the air contribution is not negligible for PBGFs, due to the high field confinement into the hollow core and the small overlap with silica. Bjarklev et al. [4.47] have shown that, in this case, the variation $\delta n_{\text{eff}}$ in Eq. (4.5) is given by

$$\begin{aligned}\delta n_{\text{eff}} &= P\left( n_{2_{\text{air}}} \frac{n_{\text{air}}^2 \epsilon_0^2 c^2 \displaystyle\int_{S_{\text{air}}} |\overline{E}|^4 dS}{(v_g \displaystyle\int_S \overline{E}\cdot\overline{D}\, dS)^2} + \right. \\ &\left. + n_{2_{\text{SiO}_2}} \frac{n_{\text{SiO}_2}^2 \epsilon_0^2 c^2 \displaystyle\int_{S_{\text{SiO}_2}} |\overline{E}|^4 dS}{(v_g \displaystyle\int_S \overline{E}\cdot\overline{D}\, dS)^2} \right),\end{aligned} \tag{4.6}$$

where $\overline{D} = \epsilon_0 \epsilon_r \overline{E}$, $c$ is the speed of light in the vacuum and $v_g$ is the group velocity of the guided mode. It is possible to demonstrate [4.48] that the guided-mode power $P$ is related to $v_g$ according to

$$P = \frac{1}{2} \int_S \overline{E} \times \overline{H}^* \cdot \hat{z} dS = \frac{1}{2} v_g \int_S \overline{E} \cdot \overline{D} dS \ . \tag{4.7}$$

Then, by substituting Eq. (4.7) into Eq. (4.6), and the last one into Eq. (4.5), the PBGF nonlinear coefficient can be expressed as

$$\gamma = \gamma_{\text{air}} + \gamma_{\text{SiO}_2} = \frac{2\pi}{\lambda} \frac{n_{2_{\text{air}}}}{A_{\text{eff}_{\text{air}}}} + \frac{2\pi}{\lambda} \frac{n_{2_{\text{SiO}_2}}}{A_{\text{eff}_{\text{SiO}_2}}} \ , \tag{4.8}$$

where $i = \text{air}, \text{SiO}_2$ and

$$A_{\text{eff}_i} = \frac{\left( \int_{S_i} \overline{E} \times \overline{H}^* \cdot \hat{z} dS \right)^2}{n_i^2 \epsilon_0^2 c^2 \int_S |\overline{E}|^2 dS} \tag{4.9}$$

is the effective area evaluated as in [4.47]. This expression of $\gamma$ has allowed to take into account both the vectorial effects and the nonlinear-index coefficient variations in the fiber cross-section, which are otherwise neglected in the definition given in [4.45].

The spectral variation of the nonlinear coefficient for the four PBGFs A, B, C, and D, whose cross-section is shown in Fig. 2.29, is reported in Fig. 4.6. Notice that the silica contribution, that is $\gamma_{\text{SiO}_2}$, is negligible for all the fibers in the wavelength range considered, with the exception of the PBG edges, since the guided mode becomes delocalized. As a consequence, the nonlinear coefficient mainly depends on the air contribution, in particular on the $A_{\text{eff}_{\text{air}}}$ value. This is the reason why $\gamma$ significantly depends on the hollow-core size, whereas its dependence on the cladding geometric parameters is very weak. It is important to underline that for the two fibers with $R = 3\Lambda$, that is for the PBGFs B and D, the value of the nonlinear coefficient is more than two orders of magnitude lower than that of standard optical fibers. In particular, it is lower than $4 \cdot 10^{-3}$ (W·km)$^{-1}$ over a large wavelength range, wider than the low-loss region and the effectively single-mode one. Finally, notice that, in order to obtain a significant reduction of the nonlinear coefficient value for the hollow-core PBGFs, it is better to increase the mode field diameter by changing, for example, the size and the shape of the hollow core, rather than to further reduce the field–silica overlap.

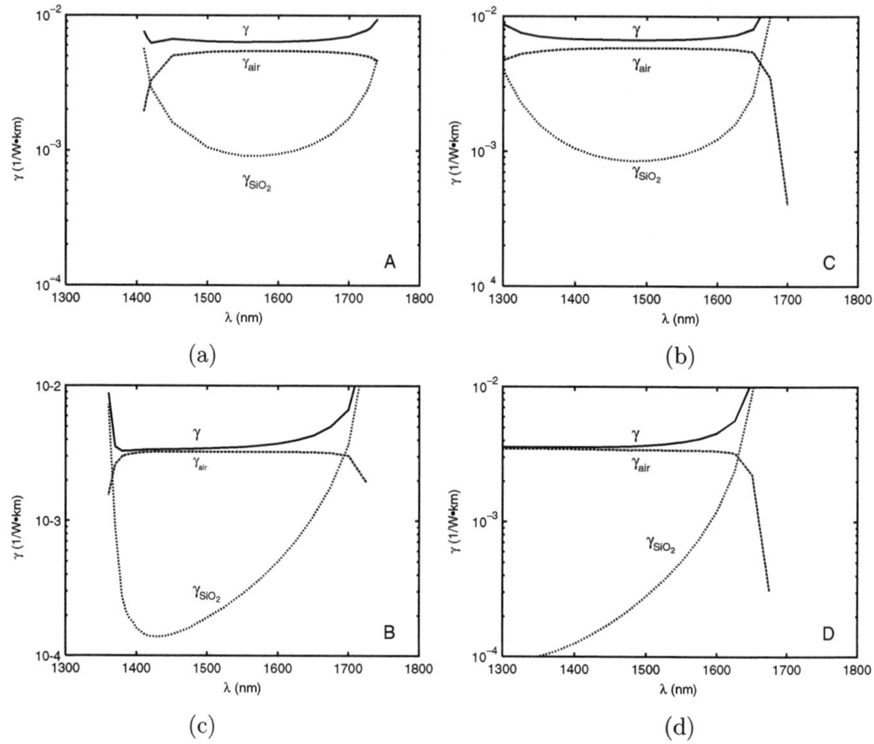


Figure **4.6**: Nonlinear coefficient versus the wavelength of the fundamental mode of the PBGFs (a) A, (b) C, (c) B, and (d) D, with $d/\Lambda = 0.6$ (*left column*) and $d/\Lambda = 0.64$ (*right column*) when the core radius is $R = 2\Lambda$ (*top*) and $R = 3\Lambda$ (*bottom*) [4.3].

# Bibliography

[4.1] F. Poli, F. Adami, M. Foroni, L. Rosa, A. Cucinotta, and S. Selleri, "Optical parametric amplification in all-silica triangular-core photonic crystal fibers," *Applied Physics B*, vol. 81, pp. 251–255, July 2005.

[4.2] S. Selleri, A. Cucinotta, F. Poli, M. Foroni, and L. Rosa, "Optical parametric amplification in dispersion-flattened highly nonlinear photonic crystal fibers," in *Proc. International Congress on Optics and Optoelectronics SPIE-COO 2005*, Warsaw, Poland, Aug. 28–Sept. 2, 2005.

[4.3] L. Vincetti, M. Maini, F. Poli, A. Cucinotta, and S. Selleri, "Numerical analysis of hollow core photonic band gap fibers with modified honeycomb lattice," *Optical and Quantum Electronics*, vol. 38, pp. 903–912, Dec. 2006.

[4.4] R. R. Alfano and S. L. Shapiro, "Observation of self-phase modulation and small-scale filaments in crytal and glasses," *Physical Review Letters*, vol. 24, pp. 592–594, Mar. 1970.

[4.5] N. I. Nikolov, T. Sørensen, O. Bang, and A. Bjarklev, "Improving efficiency of supercontinuum generation in photonic crystal fibers by direct degenerate four-wave mixing," *Journal of Optical Society of America B*, vol. 20, pp. 2329–2337, Nov. 2003.

[4.6] W. J. Wadsworth, A. Ortigosa-Blanch, J. C. Knight, T. A. Birks, T. P. M. Man, and P. St. J. Russell, "Supercontinuum generation in photonic crystal fibers and optical fiber tapers: a novel light source," *Journal of Optical Society of America B*, vol. 19, pp. 2148–2155, Sept. 2002.

[4.7] K. P. Hansen and R. E. Kristiansen, "Supercontinuum Generation in Photonic Crystal Fibers," Crystal Fibre A/S, Tech. Rep., 2005.

[4.8] K. M. Hilligsøe, H. N. Paulsen, J. Thøgersen, S. R. Keiding, and J. J. Larsen, "Initial steps of supercontinuum generation in photonic crystal fibers," *Journal of Optical Society of America B*, vol. 20, pp. 1887–1893, Sept. 2003.

[4.9] A. V. Husakou and J. Herrmann, "Supercontinuum generation, four-wave mixing, and fission of higher-order solitons in photonic-crystal fibers," *Journal of Optical Society of America B*, vol. 19, pp. 2171–2182, Sept. 2002.

[4.10] A. Apolonski, B. Povazay, A. Unterhuber, W. Drexler, W. J. Wadsworth, J. C. Knight, and P. St. J. Russell, "Spectral shaping of supercontinuum in a cobweb photonic-crystal fiber with sub-20-$fs$ pulses," *Journal of Optical Society of America B*, vol. 19, pp. 2165–2170, Sept. 2002.

[4.11] J. C. Knight, J. Arriaga, T. A. Birks, A. Ortigosa-Blanch, W. J. Wadsworth, and P. St. J. Russell, "Anomalous dispersion in photonic crystal fiber," *IEEE Photonics Technology Letters*, vol. 12, pp. 807–809, July 2000.

[4.12] A. Cucinotta, S. Selleri, L. Vincetti, and M. Zoboli, "Holey fiber analysis through the finite-element method," *IEEE Photonics Technology Letters*, vol. 14, pp. 1530–1532, Nov. 2002.

[4.13] J. K. Ranka, R. S. Windeler, and A. J. Stentz, "Visible continuum generation in air-silica microstructure optical fibers with anomalous dispersion at 800 nm," *Optics Letters*, vol. 25, pp. 25–27, Jan. 2000.

[4.14] T. Yamamoto, H. Kubota, S. Kawanishi, M. Tanaka, and S. Yamaguchi, "Supercontinuum generation at 1.55 μm in a dispersion-flattened polarization-maintaining photonic crystal fiber," *Optics Express*, vol. 11, pp. 1537–1540, June 2003. Available at: http://www.opticsexpress.org/abstract.cfm?URI=oe-11-13-1537

[4.15] M. Lehtonen, G. Genty, H. Ludvigsen, and M. Kaivola, "Supercontinuum generation in a highly birefringent microstructured fiber," *Applied Physics Letters*, vol. 82, pp. 2197–2199, Apr. 2003.

[4.16] A. Proulx, J. M. Ménard, N. Hô, J. Laniel, R. Vallée, and C. Paré, "Intensity and polarization dependences of the supercontinuum generation in birefringent and highly nonlinear microstructured fibers," *Optics Express*, vol. 11, pp. 3338–3345, Dec. 2003. Available at: http://www.opticsexpress.org/abstract.cfm?URI=oe-11-25-3338

[4.17] A. B. Fedotov, A. N. Naumov, A. M. Zheltikov, I. Bugar, D. Chorvat Jr., D. Chorvat, A. P. Tarasevitch, and D. von der Linde, "Frequency-tunable supercontinuum generation in photonic-crystal fibers by femtosecond pulses of an optical parametric amplifier," *Journal of Optical Society of America B*, vol. 19, pp. 2156–2164, Sept. 2002.

[4.18] G. Genty, M. Lehtonen, H. Ludvigsen, J. Broeng, and M. Kaivola, "Spectral broadening of femtosecond pulses into continuum radiation in microstructured fibers," *Optics Express*, vol. 10, pp. 1083–1098, Oct. 2002. Available at: http://www.opticsexpress.org/abstract.cfm?URI=oe-10-20-1083

[4.19] K. M. Hilligsøe, T. Andersen, H. Paulsen, C. Nielsen, K. Mølmer, S. Keiding, R. Kristiansen, K. Hansen, and J. Larsen, "Supercontinuum generation in a photonic crystal fiber with two zero dispersion wavelengths," *Optics Express*, vol. 12, pp. 1045–1054, Mar. 2004. Available at: http://www.opticsexpress.org/abstract.cfm?URI=oe-12-6-1045

[4.20] A. Ortigosa-Blanch, J. C. Knight, and P. St. J. Russell, "Pulse breaking and supercontinuum generation with 200-$fs$ pump pulses in photonic crystal fibers," *Journal of Optical Society of America B*, vol. 19, pp. 2567–2572, Nov. 2002.

[4.21] G. Genty, M. Lehtonen, H. Ludvigsen, and M. Kaivola, "Enhanced bandwidth of supercontinuum generated in microstructured fibers," *Optics Express*, vol. 12, pp. 3471–3480, July 2004. Available at: http://www.opticsexpress.org/abstract.cfm?URI=oe-12-15-3471

[4.22] K. P. Hansen, J. J. Larsen, J. R. Jensen, S. Keiding, J. Broeng, H. R. Simonsen, and A. Bjarklev, "Super continuum generation at 800 nm in highly nonlinear photonic crystal fibers with normal dispersion," in *Proc. Laser and Ectro-Optics Society Annual Meeting LEOS 2001*, San Diego, California, USA, Nov. 12–13, 2001, pp. 703–704.

[4.23] K. Hansen, J. R. Jensen, D. Birkedal, J. M. Hvam, and A. Bjarklev, "Pumping wavelength dependence of super continuum generation in photonic crystal fibers," in *Proc. Optical Fiber Communications Conference OFC 2002*, Anaheim, California, USA, Mar. 17–22, 2002, paper ThGG8.

[4.24] J. M. Dudley, L. Provino, N. Grossard, H. Maillotte, R. S. Windeler, B. J. Eggleton, and S. Coen, "Supercontinuum generation in air-silica microstructured fibers with nanosecond and femtosecond pulse pumping," *Journal of Optical Society of America B*, vol. 19, pp. 765–771, Apr. 2002.

[4.25] H. Hundertmark, D. Kracht, D. Wandt, C. Fallnich, V. V. R. K. Kumar, A. K. George, J. C. Knight, and P. St. J. Russell, "Supercontinuum generation with 200 pJ laser pulses in an extruded SF6 fiber at 1560 nm," *Optics Express*, vol. 11, pp. 3196–3201, Dec. 2003. Available at: http://www.opticsexpress.org/abstract.cfm?URI=oe-11-24-3196

[4.26] S. Coen, A. H. L. Chau, R. Leonhardt, J. D. Harvey, J. C. Knight, W. J. Wadsworth, and P. St. J. Russell, "White-light supercontinuum generation with 60-ps pump pulses in a photonic crystal fiber," *Optics Letters*, vol. 26, pp. 1356–1358, Sept. 2001.

[4.27] *Nonlinear photonic crystal fibers – Selected datasheets – 1550 nm fibers*, Crystal Fibre A/S. Available at: http://www.crystal-fibre.com/datasheets/NL-1550-POS-1.pdf

[4.28] "Application note on supercontinuum in SC-5.0-1040," Crystal Fibre A/S, Tech. Rep., 2005. Available at: http://www.crystal-fibre.com/support/Supercontinuum-SC-5.0-1040.pdf

[4.29] I. Hartl, X. D. Li, C. Chudoba, R. K. Ghanta, T. H. Ko, J. G. Fujimoto, J. K. Ranka, and R. S. Windeler, "Ultrahigh-resolution optical coherence tomography using continuum generation in an air-silica microstructure optical fiber," *Optics Letters*, vol. 26, pp. 608–610, May 2001.

[4.30] J. Hansryd, P. A. Anderkson, M. Westlund, J. Li, and P. Hedekvist, "Fiber-based optical parametric amplifiers and their applications," *IEEE Journal of Selected Topics in Quantum Electronics*, vol. 8, pp. 506–520, May/June 2002.

[4.31] R. H. Stolen and J. E. Bjorkholm, "Parametric amplification and frequency conversion in optical fiber," *IEEE Journal of Quantum Electronics*, vol. QE-18, pp. 1062–1072, July 1982.

[4.32] G. Cappellini and S. Trillo, "Third order three-wave mixing in single mode fibers: exact solutions and spatial instability effects," *Journal of Optical Society of America B*, vol. 8, pp. 824–838, Apr. 1991.

[4.33] C. J. McKinstrie, S. Radic, and A. R. Chraplyvy, "Parametric amplifiers driven by two pump waves," *IEEE Journal of Selected Topics in Quantum Electronics*, vol. 8, pp. 538–547, May/June 2002.

[4.34] J. Hansryd and P. A. Andrekson, "Broad-band continuous-wave-pumped fiber optical parametric amplifier with 49 dB gain and wavelength conversion efficiency," *IEEE Photonics Technology Letters*, vol. 13, pp. 194–196, Mar. 2001.

[4.35] R. Tang, J. Lasri, P. Devgan, J. E. Sharping, and P. Kumar, "Microstructure-fibre-based optical parametric amplifier with gain slope of $\simeq$200 dB/W/km in the telecom range," *Electronics Letters*, vol. 39, pp. 195–196, Jan. 2003.

[4.36] W. Belardi, J. H. Lee, K. Furusawa, A. Yusoff, P. Petropoulos, M. Ibsen, T. M. Monro, and D. J. Richardson, "A 10 Gbit/s tunable wavelength converter based on four-wave-mixing in higly nonlinear holey fiber," in *Proc. European Conference on Optical Communication ECOC 2002*, Copenhagen, Denmark, Sept. 8–12, 2002, paper postdeadline 1.2.

[4.37] J. D. Harvey, R. Leonhardt, S. Coen, G. K. L. Wong, J. C. Knight, W. J. Wadsworth, and P. St. J. Russell, "Scalar modulation instability in the normal dispersion regime by use of a photonic crystal fiber," *Optics Letters*, vol. 28, pp. 2225–2227, Nov. 2003.

[4.38] T. V. Andersen, K. M. Hilligsøe, C. K. Nielsen, J. Thøgersen, K. P. Hansen, S. R. Keiding, and J. J. Larsen, "Continuous-wave wavelength conversion in a photonic crystal fiber with two zero-dispersion wavelengths," *Optics Express*, vol. 12, pp. 4113–4122, Aug. 2004. Available at: http://www.opticsinfobase.org/abstract.cfm?URI=oe-12-17-4113

[4.39] C. J. S. de Matos, J. R. Taylor, and K. P. Hansen, "Continuous-wave, totally fiber integrated optical parametric oscillator using holey fiber," *Optics Letters*, vol. 29, pp. 983–985, May 2004.

[4.40] V. Finazzi, T. M. Monro, and D. J. Richardson, "Small-core silica holey fibers: nonlinearity and confinement loss trade-offs," *Journal of Optical Society of America B*, vol. 20, pp. 1427–1436, July 2003.

[4.41] G. Renversez, B. Kuhlmey, and R. McPhedran, "Dispersion management with microstructured optical fibers: ultraflattened chromatic dispersion with low losses," *Optics Letters*, vol. 28, pp. 989–991, June 2003.

[4.42] F. Poli, A. Cucinotta, S. Selleri, and A. H. Bouk, "Tailoring of flattened dispersion in highly nonlinear photonic crystal fibers," *IEEE Photonics Technology Letters*, vol. 16, pp. 1065–1067, Apr. 2004.

[4.43] D. J. Richardson, F. Poletti, J. Y. Y. Leong, X. Feng, H. Ebendorff-Heidepriem, V. Finazzi, K. E. Frampton, S. Asimakis, R. C. Moore,

J. C. Baggett, J. R. Hayes, M. N. Petrovich, M. L. Tse, R. Amezcua, J. H. V. Price, N. G. R. Broderick, P. Petropoulos, and T. M. Monro, "Advances in microstructured fiber technology," in *Proc. IEEE/LEOS Workshop on Fibres and Optical Passive Components WFOPC 2005*, Palermo, Italy, June 22–24, 2005.

[4.44] L. Vincetti, F. Poli, and S. Selleri, "Confinement loss and nonlinearity analysis of air-guiding modified honeycomb photonic bandgap fibers," *IEEE Photonics Technology Letters*, vol. 18, pp. 508–510, Feb. 2006.

[4.45] G. P. Agrawal, *Nonlinear Fiber Optics.* New York: Academic, 2001.

[4.46] D. G. Ouzounov, F. R. Ahmad, D. Müller, N. Venkataraman, M. T. Gallagher, M. G. Thomas, J. Silcox, K. W. Koch, and A. L. Gaeta, "Generation of megawatt optical solitons in hollow-core photonic band-gap fibers," *Science*, vol. 301, pp. 1702–1704, Sept. 2003.

[4.47] J. Lægsgaard, N. A. Mortensen, and A. Bjarklev, "Mode areas and field-energy distribution in honeycomb photonic bandgap fibers," *Journal of Optical Society of America B*, vol. 20, pp. 2037–2045, Oct. 2003.

[4.48] A. W. Snyder and J. D. Love, *Optical Waveguide Theory.* New York: Chapman & Hall, 1996.

# Chapter 5

# Raman properties

The importance of Raman amplification in optical communication systems has become more and more relevant in the last years. The gain mechanism in Raman amplification is the SRS, that is, a nonlinear scattering process by which energy is transferred from a pump wavelength to a signal one, which can be longer, in the Stokes process, or shorter, in the anti-Stokes one [5.1, 5.2]. The gain flexibility, that is, the possibility to obtain amplification at any wavelength in any fiber, is one of the key advantage of Raman fiber amplifiers. Broadband and low noise-figure Raman amplifiers can be obtained with multipumping schemes [5.3–5.5]. Moreover, distributed Raman amplification provides a significant improvement of the noise performances and an increase of the signal power budget in transmission fibers [5.6, 5.7].

As already observed in previous Chapters, PCFs can greatly enhance nonlinear effects [5.8–5.11], compared to conventional optical fibers. As a consequence, PCFs can be successfully used as Raman amplification fibers [5.12]. A continuous-wave pumped Raman laser [5.13], as well as an $L^+$-band Raman amplifier in a PCF [5.14] have been already experimentally demonstrated. Enhanced Raman properties can be obtained in both index-guiding and hollow-core PCFs. For example, by filling the PCF hollow-core with hydrogen, it is possible to reduce the threshold power for the SRS by two order of magnitude [5.15]. On the contrary, the nonlinear properties of index-guiding fibers can be improved by changing only the geometric characteristics of the air-hole lattice in the PCF cross-section, or by introducing a proper germania-doped area in the fiber core.

Two meaningful parameters, that is the Raman effective area $A_{\mathrm{eff}}^{\mathrm{R}}$ and the Raman gain coefficient $\gamma_{\mathrm{R}}$, have been considered to describe the fiber Raman performances. The Raman effective area [5.16] takes into account the overlap between the field profiles of the pump and the signal, which participate to the Raman amplification process, thus providing a more complete description of the Raman properties of the fibers.

Triangular PCFs have been considered in order to design nonlinear fibers with enhanced performances for Raman amplification. The behavior of $A_{\mathrm{eff}}^{\mathrm{R}}$ and $\gamma_{\mathrm{R}}$ as a function of the triangular lattice geometric parameters, that is $\Lambda$ and $d/\Lambda$, has been investigated for all-silica PCFs and germania-doped core ones [5.16, 5.17]. Results of the present analysis have shown that a proper design of triangular PCFs can significantly improve the Raman gain performances, that is minimize the Raman effective area and maximize the Raman gain coefficient. Germania-doped triangular PCFs have been analyzed too, showing that the best Raman gain coefficient value can be obtained when the doped area is internally tangent to the first air-hole ring in the fiber cross-section. Moreover, the Raman properties of tellurite-based triangular PCFs have been evaluated and compared with the silica-based ones. Simulation results have demonstrated that, by fixing the geometric parameters and by changing only the glass matrix from silica to tellurite, an increase of two order of magnitude in the triangular PCF Raman gain coefficient is possible, due to the better Raman properties of the tellurite glass [5.18]. Triangular PCFs with enlarging air-holes and a germania-doped core have been also considered, in order to decrease the coupling losses to the standard single-mode fibers. In fact, all the PCFs with good Raman properties, that is with enhanced nonlinearity, have a small core diameter. The design analysis here reported has provided useful informations for a proper trade-off between the effective area and the Raman gain coefficient values, in order to successfully employ highly nonlinear PCFs for actual applications [5.19].

After investigating the Raman properties of solid-core triangular PCFs with different geometric characteristics, the guiding mechanism based on the PBG has been considered to design honeycomb PCFs with a germania-doped solid core, which present enhanced Raman properties. Results have demonstrated that, with respect to silica-core honeycomb PCFs, the proposed fibers with a germania-doped core avoid a drastic reduction of the effective area, while providing considerable higher Raman gain coefficient values [5.20]. Different germania concentrations has been considered too.

The gain and noise performances of different triangular PCF-based Raman amplifiers have been also analyzed. Results have shown that good Raman gain performances can be obtained by changing the PCF geometric characteristics and the germania concentration [5.21]. Moreover, simulations have been performed with low-loss triangular PCFs recently fabricated. It has been demonstrated that the maximum Raman gain achievable is strongly influenced by the fiber background losses, which are particularly high in nonlinear PCFs, especially at the pump wavelength.

Starting from the last consideration, the performances of Raman amplifiers based on triangular PCFs have been analyzed by evaluating the potential improvements obtainable with a reduction of the background losses. Simulation results have shown that the Raman amplifier performances are strongly affected by the attenuation, in particular by the difference of the loss level at the pump and the signal wavelengths [5.22].

A further analysis has been performed on Raman amplifiers based on low-loss triangular PCFs when multiple pumps are used. Different pump wavelengths and power distributions have been considered with the aim to reduce the gain spectrum ripple as much as possible. The study here presented has demonstrated that a flat Raman gain can be obtained in both the C and L band [5.23, 5.24]. Higher gain values can be reached in the latter band, since the attenuation at the pump wavelengths used in this case is lower, while the efficiency of the pumps for the C band is strongly reduced by the OH-peak attenuation.

All the results here summarized clearly show that, in order to completely exploit the good PCF Raman properties, it is fundamental to develop a fully optimized PCF fabrication process, necessary to reduce not only the total attenuation, but also the OH-absorption peak, which strongly affects the pump efficiency.

## 5.1 Raman effective area and Raman gain coefficient

The Raman propagation equations for one signal and one pump interacting, in the continuous-wave case and neglecting the double Rayleigh backscattering and the amplified spontaneous Raman scattering, are

$$\frac{dI_s}{dz} = g_R I_s I_p - \alpha_s I_s \tag{5.1}$$

$$\frac{dI_p}{dz} = \pm\left(\frac{\omega_p}{\omega_s} g_R I_p I_s + \alpha_p I_p\right) , \tag{5.2}$$

where $I_s$ and $I_p$ are the signal and the pump intensities, $\alpha_s$ and $\alpha_p$ account for the fiber losses at the signal and the pump frequency, respectively, that is $\omega_s$ and $\omega_p$, and $g_R$ is the Raman gain efficiency between the pump and the signal [5.25]. By integrating Eqs. (5.1) and (5.2) on the fiber transverse section $S$, it yields

$$\frac{dP_s}{dz} = \gamma_R P_s P_p - \alpha_s P_s \tag{5.3}$$

$$\frac{dP_p}{dz} = \pm\left(\hat{\gamma}_R P_p P_s + \alpha_p P_p\right) , \tag{5.4}$$

where $P_s$ and $P_p$ are the signal and the pump powers and $\gamma_R$ is the Raman gain coefficient, defined as

$$\gamma_R = \frac{\iint_S g_R(x,y) I_s(x,y) I_p(x,y)\, dx\, dy}{\iint_S I_s(x,y)\, dx\, dy \iint_S I_p(x,y)\, dx\, dy} . \tag{5.5}$$

The coefficient

$$\hat{\gamma}_R = \frac{\omega_p}{\omega_s}\gamma_R \tag{5.6}$$

has been also introduced in Eq. (5.4).

By taking into account that

$$P_k = \iint_S I_k(x,y)\, dx\, dy \quad k = \mathrm{s}, \mathrm{p} , \tag{5.7}$$

it is possible to provide an alternative definition of the Raman gain coefficient starting from Eq. (5.5), that is,

$$\gamma_R = \iint_S g_R(x,y) i_s(x,y) i_p(x,y)\, dx\, dy ; \tag{5.8}$$

where $i_s$ and $i_p$ are the signal and the pump normalized intensities, respectively, which satisfy the following conditions

$$\iint_S i_k(x,y)\, dx\, dy = 1 \tag{5.9}$$

and

$$i_k(x,y) = \frac{I_k(x,y)}{P_k} \quad k = \mathrm{s},\mathrm{p} \ . \tag{5.10}$$

Moreover, since the Raman properties strongly depend on the medium, a precise evaluation of the Raman gain coefficient requires the knowledge of the different contributions given by the materials which constitute the fiber. Since the most common fibers consist of a silica host, $SiO_2$, with germania, $GeO_2$, added in the core to increase the refractive index, the definition of the Raman gain coefficient can be extended as in [5.26], that is,

$$\begin{aligned} \gamma_{\mathrm{R}} &= \iint_S C_{\mathrm{SiSi}}(1-2m(x,y))i_{\mathrm{s}}(x,y)i_{\mathrm{p}}(x,y)dxdy \\ &+ \iint_S C_{\mathrm{GeSi}}2m(x,y)i_{\mathrm{s}}(x,y)i_{\mathrm{p}}(x,y)dxdy \ . \end{aligned} \tag{5.11}$$

In this expression $\gamma_{\mathrm{R}}$ has been decomposed into a sum of two contributions, one from pure silica, with bound Si–O–Si, and the other from binary germania and silica, with bound Ge–O–Si. Notice that each contribution is calculated by integrating the normalized intensities of the pump and the signal, weighted by the fractional distribution of the bridges Si–O–Si, that is $(1-2\,m(x,y))$, and Ge–O–Si, that is $2\,m(x,y)$, being $m(x,y)$ the germania concentration [5.26]. Equation (5.11) is implicitly dependent on the spectral separation $\Delta\nu$ between the pump and the signal through $C_{\mathrm{SiSi}}(\Delta\nu)$ and $C_{\mathrm{GeSi}}(\Delta\nu)$, which are the Raman spectra relative to the bounds Si–O–Si and Ge–O–Si, respectively. However, in the following analysis a fixed separation between the pump and the signal $\Delta\nu \simeq 13.2$ THz has been assumed, which corresponds to a wavelength separation of about 100 nm in the C band, in order to consider the peak Raman gain coefficient. The pump and the signal wavelengths have been chosen equal to 1450 and 1550 nm, respectively, but the same analysis can be performed for any wavelengths of the interacting signals. The peak gain spectra $C_{\mathrm{SiSi}}$ and $C_{\mathrm{GeSi}}$ in Eq. (5.11) have been evaluated following the method presented in [5.26]. By taking into account the results for pairs of different fibers with known Raman gain coefficient values, it has been found that $C_{\mathrm{SiSi}} = 3.34 \cdot 10^{-14}$ m/W and $C_{\mathrm{GeSi}} = 1.18 \cdot 10^{-13}$ m/W.

The Raman effective area can be defined as it follows

$$\begin{aligned} A_{\mathrm{eff}}^{\mathrm{R}} &= \frac{\iint_S I_{\mathrm{s}}(x,y)\,dx\,dy \iint_S I_{\mathrm{p}}(x,y)\,dx\,dy}{\iint_S I_{\mathrm{s}}(x,y)I_{\mathrm{p}}(x,y)\,dx\,dy} \\ &= \frac{1}{\iint_S i_{\mathrm{s}}(x,y)i_{\mathrm{p}}(x,y)\,dx\,dy} \ . \end{aligned} \tag{5.12}$$

Considering a mean value of the Raman gain efficiency $\overline{g}_{\mathrm{R}}$ in the fiber cross-section, the relation between the Raman gain coefficient and the Raman effective area can be expressed as

$$\gamma_{\mathrm{R}} = \frac{\overline{g}_{\mathrm{R}}}{A_{\mathrm{eff}}^{\mathrm{R}}} \; . \tag{5.13}$$

Notice that the coefficient $\overline{g}_{\mathrm{R}}$ represents a total value of the Raman gain efficiency associated with the fiber, which takes into account the materials that compose the fiber and their spatial distribution. If the interacting signals have the same frequency, the Raman effective area coincides with that given by the "classical" definition

$$A_{\mathrm{eff}} = \frac{\left(\iint_S I(x,y)\, dx\, dy\right)^2}{\iint_S I(x,y)^2\, dx\, dy} \; . \tag{5.14}$$

According to the previous definition of Eq. (5.12), the Raman effective area usually presents values between those calculated with the expression in Eq. (5.14) at the pump and the signal wavelength. In fact, $A_{\mathrm{eff}}^{\mathrm{R}}$ accounts for the overlap between both the fields on the fiber cross-section, thus providing an insight into the strength of the Raman interaction. The Raman effective area contains implicitly more information than the "classical" $A_{\mathrm{eff}}$, since it is a function of the fiber geometry and of the signal wavelength, but also of the pump wavelength or, equivalently, of the frequency separation between the pump and the signal. For these reasons it is a more complete parameter for the description of the Raman properties.

The normalized intensities of the pump and the signal which appear in the previous equations have been evaluated as described in Appendix A. In particular, the intensity has been calculated according to the Poynting vector definition reported in Eq. (A.5), which has a general validity. This formulation is more accurate than the one presented in [5.27]. Moreover, it is important to underline that there is a difference in the calculation of the Raman effective area and the Raman gain coefficient. In fact, only the glass zones in the PCF cross-section give a contribution to the Raman gain coefficient, while in the calculation of the effective area it is necessary to consider all the fiber section, that is also the contribution of the field in the air-holes.

## 5.2 Raman properties of triangular PCFs

The Raman properties of triangular PCFs have been investigated by changing the geometric parameters that characterize the fiber cross-section, that is the air-filling fraction $d/\Lambda$ and the pitch $\Lambda$ of the air-hole lattice, as well as the characteristics of the fiber solid core. In fact, by properly modifying these parameters, it is possible to change the guided-mode field distribution and, as a consequence, the Raman effective area and the Raman gain coefficient.

### 5.2.1 Silica triangular PCFs

Initially, PCFs with a triangular lattice of air-holes in a silica bulk, shown in Fig. 3.1 with the geometric parameters $d$ and $\Lambda$, have been considered for the Raman property analysis. In order to obtain fibers with good nonlinear properties, triangular PCFs with large air-holes and small pitch have been designed. In particular, $d/\Lambda$ has been chosen equal to 0.6, 0.7, 0.8, and 0.9, and a pitch which varies between 0.7 and 2.3 μm has been considered. Notice that the number of air-hole rings is variable, since it depends on the geometric parameters of the studied PCFs. However, simulation results have demonstrated that eight air-hole rings are enough for all the studied PCFs to obtain accurate values of the Raman effective area and the Raman gain coefficient.

Figure 5.1a and b report the values of the Raman effective area and the Raman gain coefficient, respectively, as a function of the pitch $\Lambda$ for the

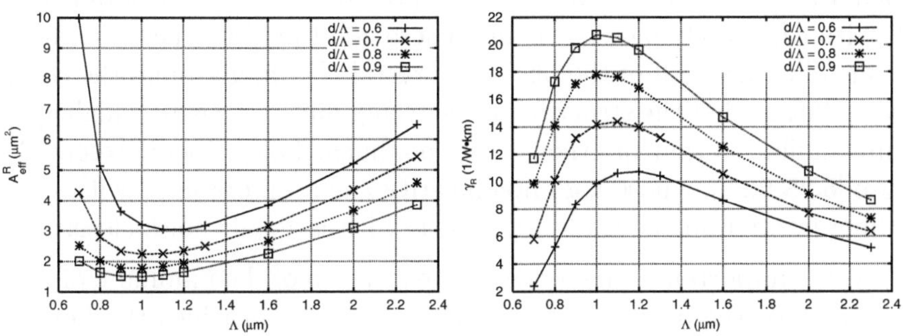


(a) (b)

Figure **5.1**: (a) Raman effective area and (b) Raman gain coefficient behavior as a function of the pitch $\Lambda$ for the PCFs with different $d/\Lambda$, that is 0.6, 0.7, 0.8, and 0.9 [5.16].

triangular PCFs with $d/\Lambda$ in the range 0.6–0.9. Results have shown that, for a fixed $d/\Lambda$, there is an optimum value of $\Lambda$ which minimizes $A_{\text{eff}}^{\text{R}}$ and maximizes $\gamma_{\text{R}}$. This condition is achieved when the PCF presents a high refractive index difference between the silica core and the photonic crystal cladding and, at the same time, a small core radius, thus providing a high guided-mode field confinement. It is important to underline that in a PCF with a given $d/\Lambda$ the core dimension increases by enlarging the pitch $\Lambda$. On the contrary, the average refractive index of the microstructured cladding remains the same, since it depends only on the air-filling fraction. In order to explain the $A_{\text{eff}}^{\text{R}}$ behavior versus the pitch, it is useful to consider the two extreme situations, that is $\Lambda \to 0$ and $\Lambda \to \infty$. Being $d/\Lambda$ constant, in the first case the core radius of the triangular PCF tends to disappear, that is $\Lambda - d/2 \to 0$. As a consequence, the field is no more guided and $A_{\text{eff}}^{\text{R}} \to \infty$. The second case is quite similar, since it corresponds to a PCF with air-holes of infinite dimension, infinitely separated. The core radius $\Lambda - d/2 \to \infty$ and again $A_{\text{eff}}^{\text{R}} \to \infty$. For pitch values between these two extreme conditions, the field of the guided mode is confined in the PCF silica core, and $A_{\text{eff}}^{\text{R}}$ reaches a minimum for a well-defined optimum $\Lambda$. Analogous conclusions can be drawn for the Raman gain coefficient. In fact, it results inversely proportional to the Raman effective area in Eq. (5.11) when the fiber is undoped, that is $m = 0$. Hence, a minimum of the Raman effective area corresponds to a maximum of the Raman gain coefficient.

Moreover, Fig. 5.1 shows that the PCFs with $d/\Lambda = 0.9$ have the smallest Raman effective area and the highest values of the Raman gain coefficient. In particular, a maximum $\gamma_{\text{R}}$ of about 21 $(\text{W}\cdot\text{km})^{-1}$ has been obtained for the PCF with $d/\Lambda = 0.9$ and $\Lambda = 1$ μm. These good performances are due to the high air-filling fraction of the photonic crystal cladding around the central silica core. Notice that the structure of the triangular PCFs with a high $d/\Lambda$ value is quite similar to that of other highly nonlinear holey fibers presented in literature [5.28, 5.29], whose high nonlinearity has been obtained with very large air-filling fraction in the cladding, that is with large air-holes closely spaced.

Another important issue concerns the location of the minimum of $A_{\text{eff}}^{\text{R}}$, that is the value of the optimum $\Lambda$, called $\Lambda_{\text{opt}}$, which provides the best Raman effective area. In Table 5.1 the geometric parameters of the PCFs with the best Raman performances are summarized. In particular, the values of the PCF core radius $r_{\text{co}}$, assumed equal to $\Lambda - d/2$, are reported in the third column, while the product between the relative air-hole diameter $d/\Lambda$ and the core radius is shown in the fourth column. It is interesting to underline that all

Table 5.1: Parameters of the PCFs which provide the best Raman performances [5.16].

| $d/\Lambda$ | $\Lambda_{opt}$ (µm) | $r_{co}$ (µm) | $d/\Lambda \cdot r_{co}$ (µm) |
|---|---|---|---|
| 0.6 | 1.2 | 0.84 | 0.504 |
| 0.7 | 1.1 | 0.715 | 0.505 |
| 0.8 | 1 | 0.6 | 0.48 |
| 0.9 | 1 | 0.55 | 0.495 |

the PCFs with the best performances are characterized by a product $d/\Lambda \cdot r_{co}$ around 0.5 µm. Since $d/\Lambda$ describes the PCF air-filling fraction, it is strictly related to the refractive index difference $\Delta n$ between the core and the photonic crystal cladding. Results reported in Table 5.1 show that all the PCFs with the minimum $A_{eff}^{R}$ have a well defined combination of $\Delta n$ and $r_{co}$. This relation is really meaningful, because it allows to calculate a first approximation of $\Lambda_{opt}$ for a fixed $d/\Lambda$.

Afterward, it has been also investigated the possibility of improving the Raman amplification performances of triangular PCFs by introducing a germania-doped area in the fiber core, by exploiting the better Raman properties of germania with respect to silica. This study has been developed by following two different directions. Initially, it has been investigated how the Raman gain coefficient changes when the germania concentration in the doped core area increases, while all the other geometric parameters are kept constant. PCFs with $\Lambda = 1.6$ µm, $d/\Lambda$ equal to 0.4, 0.6, and 0.8, and a circular $GeO_2$-doped region with radius $R_d = \Lambda/2$ have been considered. The germania concentration has been progressively increased, starting from 5% up to 25% mol, with step of 5% mol. An example of the analyzed PCFs, that is the one with $d/\Lambda = 0.6$ and $\Lambda = 1.6$ µm, is shown in Fig. 5.2a. The evaluated values of the Raman gain coefficient are reported in Fig. 5.2b as a function of the germania concentration. Notice that, for a fixed $d/\Lambda$, $\gamma_R$ linearly increases with the $GeO_2$ concentration. Moreover, an excellent agreement has been obtained with the linear interpolation of the $\gamma_R$ values measured by Galeener *et al.* [5.30] or the ones reported in [5.31].

A second analysis has been carried out, in order to find the optimum germania-doped region dimension. Once again PCFs with $d/\Lambda$ equal to 0.6, 0.7, 0.8, and 0.9 have been considered. For each $d/\Lambda$, it has been chosen a pitch $\Lambda = \Lambda_{opt}$, previously calculated for the all-silica PCFs and reported in

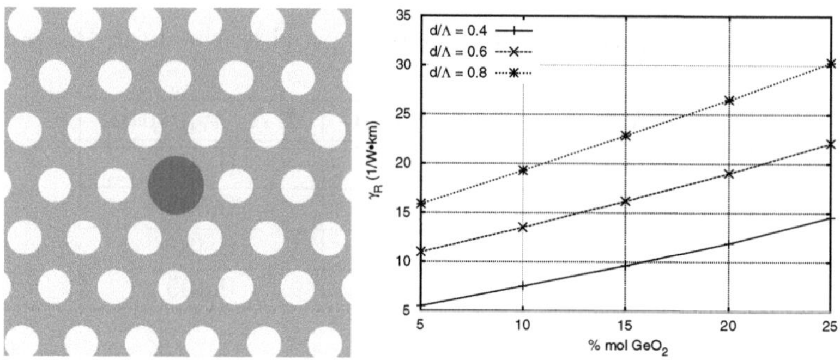


(a) (b)

Figure **5.2**: (a) Transverse section of the germania-doped PCF with $d/\Lambda = 0.6$, $\Lambda = 1.6$ µm and $R_d = \Lambda/2$. (b) Raman gain coefficient of the germania-doped PCFs with $\Lambda = 1.6$ µm, $R_d = \Lambda/2$ and $d/\Lambda = 0.4$, 0.6, and 0.8, as a function of the germania concentration [5.16].

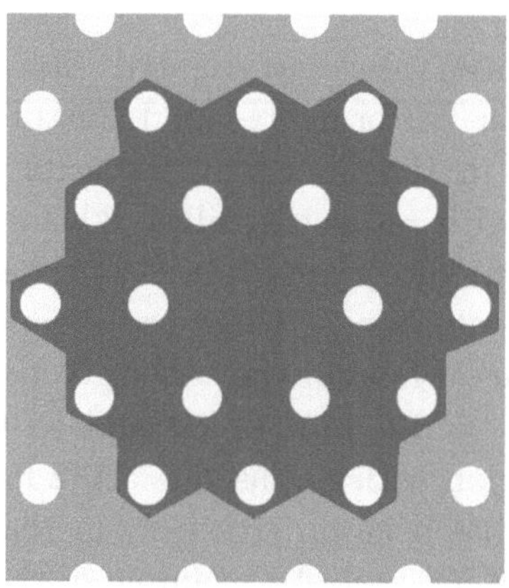

Figure **5.3**: Transverse section of the germania-doped PCF with $d/\Lambda = 0.7$, $\Lambda = 1.1$ µm and $R_d \simeq 2.75$ µm, corresponding to a doped-region over the second ring of air-holes [5.16].

Table 5.1. The $GeO_2$ concentration has been fixed to 20% mol, which corresponds to a refractive index of 1.47 at 1550 nm for the doped silica. The germania-doped area has been progressively enlarged, starting from a circular region with radius $R_d \simeq \Lambda/2$ up to an area which includes the first three air-hole rings. An example of a doped PCF considered in the present analysis, that is the one with $d/\Lambda = 0.7$ and $\Lambda = 1.1$ µm, is shown in Fig. 5.3. Notice that, when the doped region is internal to the first air-hole ring, the radius $R_\mathrm{d}$ can be straight defined. Otherwise, if the $GeO_2$-doped region extends over the first air-hole ring, as in the PCF reported in Fig. 5.3, only a mean value of $R_\mathrm{d}$ can be defined. Such a doped-region profile has been chosen in order to analyze structures as much as possible similar to the physically feasible PCFs, since perfect circular doped regions extending in the photonic crystal cladding can not be easily fabricated.

Figure 5.4a and b report the values of the Raman effective area and the Raman gain coefficient of the PCFs previously described, as a function of the mean doped-area radius $R_\mathrm{d}$. It is important to underline that a maximum $\gamma_\mathrm{R}$ value exists for each PCF. This occurs when the germania-doped area is internally tangent to the first air-hole ring, that is for a well-defined value of the doped-area radius $R_\mathrm{d,opt} = \Lambda - d/2$. Since $R_\mathrm{d,opt}$ is different for the considered PCFs, being related to their geometric parameter values, the position of the maximum is not the same for all the $\gamma_\mathrm{R}$ curves. Notice that the $A_\mathrm{eff}^\mathrm{R}$ minima, reported in Fig. 5.4a, have been obtained for $R_\mathrm{d}$ values which are sometimes different from $R_\mathrm{d,opt}$. This mismatch can be explained considering that the Raman gain coefficient of a germania-doped PCF is calculated according to Eq. (5.11). As a consequence, $\gamma_\mathrm{R}$ is not simply inversely proportional to $A_\mathrm{eff}^\mathrm{R}$, as it happens in all-silica PCFs. In fact, it depends on the guided-mode field

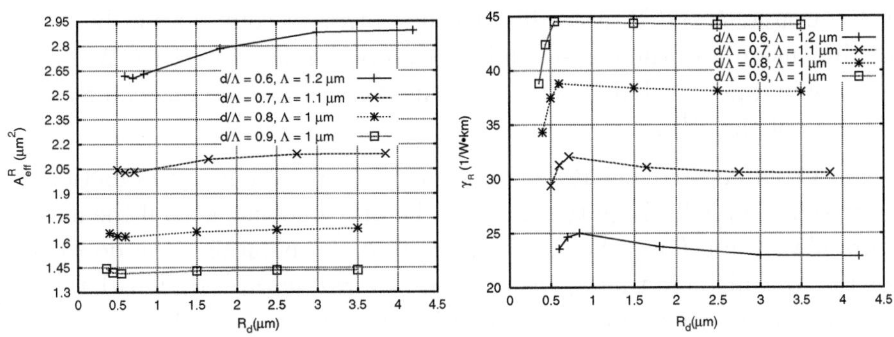


(a) (b)

Figure **5.4**: (a) Raman effective area and (b) Raman gain coefficient of the PCFs doped with 20% mol of $GeO_2$, with $d/\Lambda = 0.6$, 0.7, 0.8, and 0.9, and $\Lambda_\mathrm{opt} = 1.2$, 1.1, 1, and 1 µm, respectively, as a function of the doped-area mean radius [5.16].

confinement and also on the overlap between the field and the germania-doped region. Being $C_{\mathrm{GeSi}} > C_{\mathrm{SiSi}}$, the more the field is confined in the doped region, the higher is $\gamma_{\mathrm{R}}$. As a consequence, situations can occur when the maximum $\gamma_{\mathrm{R}}$ value does not correspond to the $A_{\mathrm{eff}}^{\mathrm{R}}$ minimum.

It is possible to explain the behavior of the Raman effective area as a function of $R_{\mathrm{d}}$. $A_{\mathrm{eff}}^{\mathrm{R}}$ depends on two factors, the refractive index difference $\Delta n$ and the core dimension. For the doped PCFs it is difficult to give a plain definition of the core area. For example, if $R_{\mathrm{d}} < \Lambda - d/2$, the core can be assumed equal to the germania-doped region, while, if $R_{\mathrm{d}} > \Lambda - d/2$, the core is approximatively the circular area internally tangent to the first air-hole ring with $r_{\mathrm{co}} = \Lambda - d/2$. When the doped region is tangent to the first air-hole ring, the field senses both the refractive index difference between the germania-doped core and the silica bulk, and that between the former and the surrounding air-holes. On the contrary, when the dopant extends over the first air-hole ring, the field confinement is mainly due to the refractive index difference between the doped core and the air-holes. As a consequence, the field tails can leak out in the bridges among the air-holes, so that the Raman effective area slightly increases with a further widening of the doped region. As reported in Fig. 5.4a, the stronger $A_{\mathrm{eff}}^{\mathrm{R}}$ variation with $R_{\mathrm{d}}$ has been obtained for the PCF with the smallest $d/\Lambda$ ratio and the largest $\Lambda$, that is the one with $d/\Lambda = 0.6$ and $\Lambda = 1.2$ μm, being larger the inter space between the neighboring air-holes. Moreover, the $A_{\mathrm{eff}}^{\mathrm{R}}$ minimum has been usually obtained for $R_{\mathrm{d}}$ around $\Lambda - d/2$, according to the optimum combination between the core dimension and the core-cladding refractive index difference $\Delta n$. In other words, only a perfect balance between the core radius and $\Delta n$ can lead to the minimum of the $A_{\mathrm{eff}}^{\mathrm{R}}$.

The explanation of the Raman gain coefficient behavior as a function of $R_{\mathrm{d}}$, reported in Fig. 5.4b, is quite similar to the previous one, even if the maximum occurs always for $R_{\mathrm{d}} = \Lambda - d/2$. Notice that, by enlarging the doped region over the first air-hole ring, $\gamma_{\mathrm{R}}$ is not significantly improved. This happens for two reasons: the Raman effective area increase when $R_{\mathrm{d}} > \Lambda - d/2$, as shown in Fig. 5.4a, and the higher field fraction falling into the air-holes, which does not contribute to the Raman gain coefficient value. On the other side, doped areas which are too small, that is with $R_{\mathrm{d}} < \Lambda - d/2$, are not useful. In fact, in the studied PCFs the guided-mode field always extends almost to the first air-hole ring, so that the wider is the doped region, obviously still with $R_{\mathrm{d}} \leq \Lambda - d/2$, the higher is $\gamma_{\mathrm{R}}$. As a consequence, the $\gamma_{\mathrm{R}}$ maximum is obtained when $R_d = \Lambda - d/2$, that is when the germania-doped region comes into contact with the first air-hole ring. This is a meaningful result, especially from a technological point of view. In fact, PCFs are fabricated by stacking tubes and rods of glass into a structure, that is a macroscopic scaled preform with the air-hole pattern required in the final fiber. Results here obtained have demonstrated that, in order to design an optimum doped PCF for Raman amplification, it is just necessary to add a central germania-doped rod in the fabricated preform, which will be the core of the drawn fiber.

Finally, it is important to point out some considerations on the absolute value of the Raman gain coefficient peak for different fiber types. By comparing the values obtained for doped triangular PCFs with those of other commercially available traditional fibers, a significant improvement can be noted. Table 5.2 reports the $\gamma_R$ peak value for a standard single-mode fiber, that is an SMF, a non-zero dispersion-shifted fiber (NZ-DSF), which is similar to a dispersion-shifted fiber (DSF), a dispersion compensating fiber, or DCF, which provides good Raman performances due to its intrinsic high nonlinearity, and, finally, a fiber designed on purpose for Raman amplification [5.32]. Although the last two fibers have high Raman gain coefficient values, their performances are lower than those reachable with the triangular PCFs here studied. In fact, PCFs can offer a tighter guided-mode field confinement, and thus an effective nonlinearity per unit length 10–100 times higher than that of traditional optical fibers [5.14]. Another interesting comparison can be made between the $\gamma_R$ of triangular PCFs and that of another highly nonlinear holey fiber with a cobweb structure of air-holes, shown in Fig. 4.1 [5.33]. Due to the large air-filling fraction surrounding the central silica core, the cobweb holey fiber is characterized by $A_{\text{eff}}^{R} \simeq 1.41$ µm$^2$ and $\gamma_R \simeq 22$ (W·km)$^{-1}$. Notice that the $\gamma_R$ values calculated for the doped triangular PCFs, whatever the $d/\Lambda$ ratio, are always higher than 20 (W·km)$^{-1}$ and, selecting a doped PCF with a high air-filling fraction, values of $\gamma_R$ higher than 40 (W·km)$^{-1}$ can be obtained, as shown in Fig. 5.4b. The enhancement can be greater if the germania concentration is increased over 20% mol or, alternatively, if other glass matrix, like tellurite, are used to fabricate the triangular PCFs [5.15, 5.34].

Table 5.2: Raman gain coefficient peak value for different fiber types [5.16].

| SMF | NZ-DSF/DSF | DCF | Raman fiber |
|---|---|---|---|
| 0.5 (W·km)$^{-1}$ | 0.7–0.8 (W·km)$^{-1}$ | 3 (W·km)$^{-1}$ | 5–6 (W·km)$^{-1}$ |

### 5.2.2 Tellurite triangular PCFs

It is interesting to investigate how the good Raman properties of tellurite glasses [5.35] can be combined with the great flexibility of PCFs [5.36]. In fact, tellurite glasses offer useful properties, such as high refractive index, good infrared transmittance, and high optical nonlinearity. Recently, it has been demonstrated that tellurite glass can be used to fabricate low-loss PCFs with the extrusion process [5.37]. In addition, strong stimulated Raman scattering has been experimentally observed in this tellurite PCF [5.37].

The Raman amplification properties of triangular PCFs have been analyzed by considering a tellurite matrix, which substitutes the silica one. The geometric parameters $\Lambda$ and $d/\Lambda$ have been changed in the same range as for the silica-based fibers. Since in tellurite-based fibers the Stokes frequency shift is about 21 THz [5.35], the pump wavelength has been fixed to 1390 nm, in order to obtain Raman amplification around 1550 nm. Moreover, the Raman gain coefficient $\gamma_R$ has been calculated according to Eq. (5.8), where $g_R$ is the Raman gain efficiency between the pump and the signal. Notice that $g_R = 0$ must be assumed in the air-holes. For the tellurite-based PCFs $g_R$ has been considered 30 times higher than that of the silica glass [5.35], whose value is $3.34 \cdot 10^{-14}$ m/W, as reported in Section 5.2.1.

Simulation results are reported in Fig. 5.5 for the PCFs with $d/\Lambda = 0.6-0.9$ and $\Lambda$ in the range 0.7–1 µm. Looking at Fig. 5.5a, it is interesting to notice

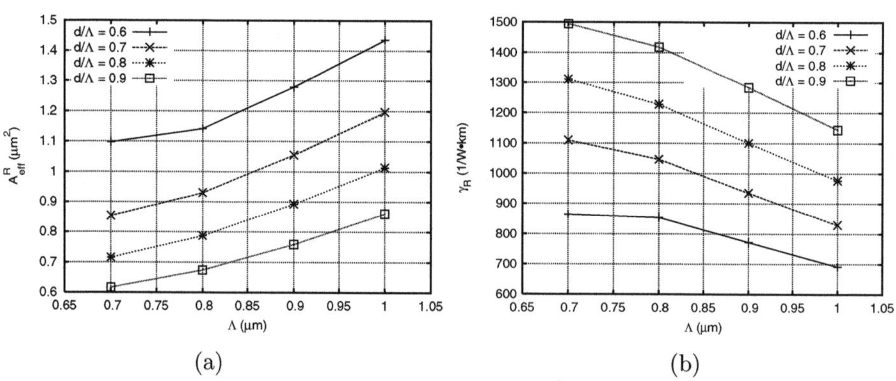


Figure **5.5**: (a) Raman effective area and (b) Raman gain coefficient values for tellurite PCFs with different $d/\Lambda$, that is 0.6, 0.7, 0.8, and 0.9, as a function of the pitch $\Lambda$ [5.18].

that the $A_{\rm eff}^{\rm R}$ is lower than in silica-based fibers and, differently from the silica case, it always increases with the pitch $\Lambda$, also in the range 0.7–1 μm. This means that both the signal and the pump fields are much more focused within the core for all the $\Lambda$ values, thanks to the higher tellurite refractive index, that is 1.998 at 1550 nm. The tellurite-based PCF with $d/\Lambda = 0.9$ and $\Lambda = 0.7$ μm has the minimum $A_{\rm eff}^{\rm R}$, that is 0.62 μm$^2$. The same fiber realized with the silica glass matrix has an $A_{\rm eff}^{\rm R}$ more than three times higher, that is about 2 μm$^2$. The lower $A_{\rm eff}^{\rm R}$ values of the tellurite-based fibers, combined with the higher Raman gain efficiency, provide a significant increase of the Raman gain coefficient. Starting from these considerations, it is possible to predict that the $\gamma_{\rm R}$ of the PCF with $d/\Lambda = 0.9$ and $\Lambda = 0.7$ μm will be nearly two order of magnitude higher than the values obtained with the silica-based PCFs. This is confirmed by the Raman gain coefficient values reported in Fig. 5.5b. It is worth saying that these extremely high $\gamma_{\rm R}$ values can allow very short PCFs to be used as Raman amplifying medium, thus reducing the impact of the higher tellurite glass background losses [5.35].

### 5.2.3 Enlarging air-hole triangular PCFs

The study of the PCF Raman performances here reported has shown that the best behavior can be obtained in triangular PCFs with a very small core dimension, which allows to obtain good nonlinear properties. As a drawback, the resulting mismatch between the field sustained by these highly nonlinear PCFs and by standard optical fibers critically increases the splicing and the coupling losses. For this reason, triangular PCFs with a fixed core diameter, still smaller than that of standard single-mode fibers, but larger than those of the PCFs investigated in Sections 5.2.1 and 5.2.2, have been considered. In addition, air-hole tailoring has been exploited to enlarge the guided-mode effective area. In order to maintain high Raman gain coefficient values, a germania-doped region has been introduced in the fiber core. However, this tightly confines the field, since the refractive index difference between the core and the photonic crystal cladding increases. As a consequence, a proper design of this kind of triangular PCFs has to provide a trade-off between the values of the Raman gain coefficient and of the effective area.

The analysis has been performed by considering three kinds of triangular PCFs with increasing dimensions of the air-holes. The hole-to-hole spacing, $\Lambda = 1$ μm, and the air-hole diameter of the first ring, $d_1 = 0.2$ μm, are the same for all the PCFs, and, consequently, also the core diameter, defined by $d_{\text{core}} = 2\Lambda - d_1$. The diameter $d_n$ of the air-holes in the $n$th outer ring, with $n \geq 1$, has been chosen according to $d_{n+1}/\Lambda = d_n/\Lambda + \Delta$. Different $\Delta$ value has been considered for the three fiber types, that is 0.05, 0.1, and 0.15, respectively. Notice that the air-hole diameters have been modified as long as their variation influences the effective area of the guided mode. The air-hole diameter in the remaining outer rings has been fixed to $d = 0.9$ μm, in order to reduce the leakage losses [5.38].

In order to obtain improved Raman performances, a germania-doped area has been introduced in all the PCFs here studied. The doped-region radius has been fixed to $r_{\text{d}} = 0.85$ μm, while the germania concentration has been changed from 0% to 19.3%. The guided-mode magnetic field at 1550 nm is reported in Fig. 5.6a for the all-silica PCF with $\Delta = 0.05$. Looking at Fig. 5.6b, it is possible to notice the higher field confinement due to the presence of a 19.3% germania-doped area in the PCF core.

The Raman performances of the triangular PCFs with enlarging air-holes have been studied by calculating the Raman gain coefficient and the Raman effective area according to Eqs. (5.11) and (5.12), respectively. The behavior of $A_{\text{eff}}^{\text{R}}$ and $\gamma_{\text{R}}$ as a function of the germania concentration for the three kinds of PCFs considered is reported in Fig. 5.7. Results have shown that, by increasing the germania concentration, the Raman effective area linearly decreases in all cases, while the Raman gain coefficient has the inverse behavior. For example, $\gamma_{\text{R}}$ is 2.1 $(\text{W}\cdot\text{km})^{-1}$ for the PCF with $\Delta = 0.05$ if only silica is present in the fiber cross-section, and it becomes about 7.2 $(\text{W}\cdot\text{km})^{-1}$ when the germania

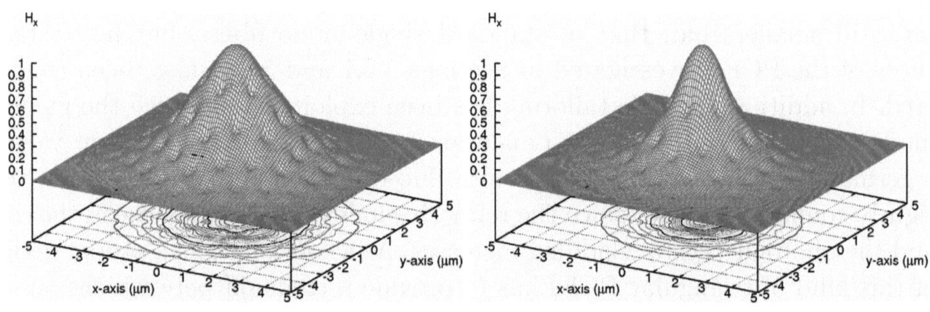


(a) (b)

Figure **5.6**: Fundamental component of the guided-mode magnetic field at 1550 nm for the (a) undoped and (b) the 19.3% germania-doped PCF with $\Delta = 0.05$.

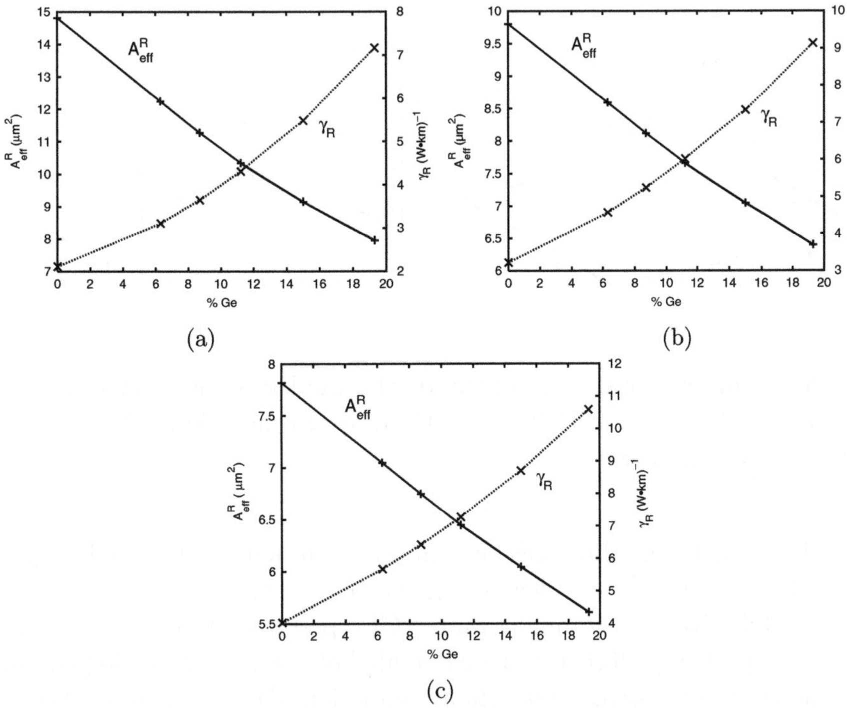


Figure **5.7**: Raman properties for the PCFs with (a) $\Delta = 0.05$, (b) $\Delta = 0.1$, and (c) $\Delta = 0.15$ with different germania concentrations.

concentration is 19.3%, as reported in Fig. 5.7a. Notice that the influence of the germania-doped area on the PCF Raman properties changes with $\Delta$, that is with the air-hole diameter in the photonic crystal cladding, even if the core diameter is the same. In fact, the Raman effective area of the PCF with $\Delta = 0.05$ is halved, by adding a germania-doped area with the highest concentration, while $A^{\mathrm{R}}_{\mathrm{eff}}$ decreases from 7.8 to 5.6 μm$^2$ for the PCF with $\Delta = 0.15$. The decrease of $A^{\mathrm{R}}_{\mathrm{eff}}$ for the PCF with the lowest $\Delta$ value can be easily understood looking at Fig. 5.6. In fact, the guided-mode field is tighter confined in the central area of the doped PCF cross-section.

## 5.3 Raman properties of honeycomb PCFs

As described in Chapter 1, PCFs guide light by two different mechanisms, that is, by modified TIR and by means of the PBG effect. The first PCF which

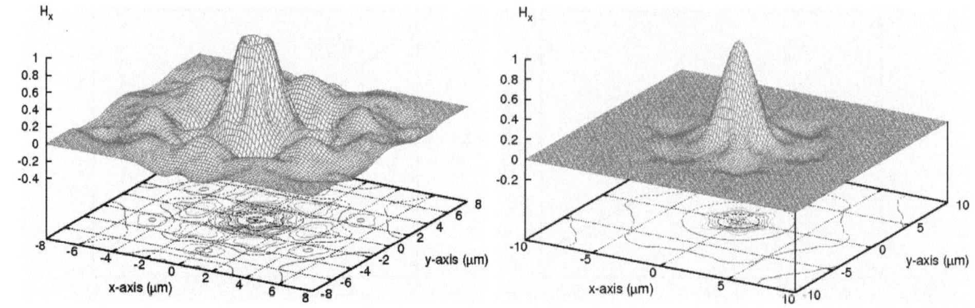


(a) (b)

Figure **5.8**: Fundamental component of the guided-mode magnetic field at 1550 nm for a honeycomb PCF (a) with an extra air-hole in the core center and (b) with a solid core.

exploited the latter guiding mechanism was characterized by a honeycomb lattice with a defect in the silica core given by an extra air-hole, as shown in Fig. 1.5 [5.39]. Even if the honeycomb PCFs present interesting and unconventional properties [5.40, 5.41], the extra air-hole causes a ring-shaped guided mode, shown in Fig. 5.8a, rather than a gaussian-like one, as in conventional fibers. Solid-core honeycomb PCFs [5.42–5.44], besides overcoming this problem, as demonstrated by the guided-mode field reported in Fig. 5.8b, open up new possibilities to further enhance the nonlinear properties, in particular the Raman ones. For this purpose, germania-doped regions can be added in the core or in the cladding of the honeycomb PCFs, provided their inclusion maintain the guided mode within the bandgap.

In the present study new solid-core honeycomb PCFs with a gaussian-like field distribution have been designed to improve the Raman gain coefficient, while assuring Raman effective area values higher than those obtained with index-guiding triangular PCFs. These requirements can be achieved by properly moving the guided solution within the bandgap through the variations of the central defect characteristics. The Raman performances of the germania-doped honeycomb PCFs have been studied again by calculating the Raman gain coefficient and the Raman effective area according to Eqs. (5.11) and (5.12), respectively.

Figure 5.9a reports the cross-section of the honeycomb PCFs here considered, which have a lattice characterized by air-holes with diameter $d = 0.5\Lambda$, and up-doped circular regions with diameter $d_{\mathrm{Ge}} = \Lambda$ and a germania

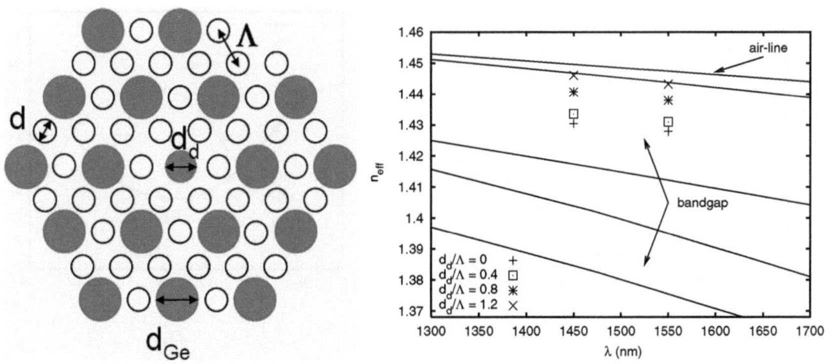


(a) (b)

Figure **5.9**: (a) Detail of the honeycomb PCF cross-section. (b) PBGs as a function of the wavelength and effective index for the pump at 1450 nm and the signal at 1550 nm versus $d_d/\Lambda$ for the PCF with $\Lambda = 2.5$ μm [5.20].

concentration of 19.3%. As shown in Fig. 5.9b for $\Lambda = 2.5$ μm, this structure presents two PBGs, which have been calculated by a freely available software package [5.45]. In order to obtain a mode guided by the PBG mechanism, a core defect has been added in the honeycomb PCF cross-section, by substituting the central area of diameter $d_{Ge}$ with a doped region of diameter $d_d$ and a lower germania concentration, that is 15%. Figure 5.9b reports also the effective index $n_{eff}$ of the guided mode at the wavelengths of the pump and the signal involved in the Raman amplification process, that is 1450 and 1550 nm, respectively, for different $d_d/\Lambda$ values. Notice that when $d_d = 0$, which corresponds to an all-silica core, the $n_{eff}$ is almost at the PBG center for both the wavelengths. As the core defect is enlarged, the effective index increases and moves toward the upper PBG edge.

This displacement of the guided state results in a lower field confinement, which counterbalances the field focusing given by the higher refractive index difference introduced by the germania-doped central region. These two opposite mechanisms acting on the field confinement allow to reduce the Raman effective area variations and, at the same time, to increase the Raman gain coefficient, thanks to the germania doping. For example, when the pitch is 2.5 μm, the two honeycomb PCFs with $d_d/\Lambda = 0$ and $d_d/\Lambda = 1.14$ have almost the same $A^R_{eff}$, but the $\gamma_R$ values are, respectively, 3.6 and 6.2 $(\text{W}\cdot\text{km})^{-1}$, as shown in Fig. 5.10a. The maximum $\gamma_R$ of 7.4 $(\text{W}\cdot\text{km})^{-1}$, which is more than twice the one of the silica-core honeycomb PCF, has been obtained when

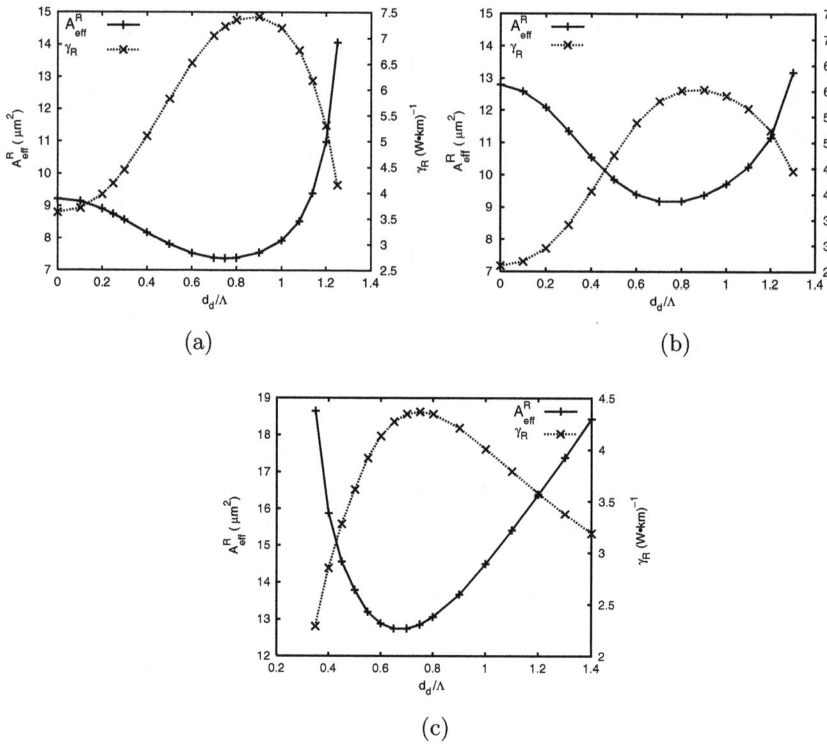


Figure **5.10**: $A_{\text{eff}}^{\text{R}}$ and $\gamma_{\text{R}}$ versus the core defect diameter normalized to the pitch $d_{\text{d}}/\Lambda$ for the honeycomb PCFs with (a) $\Lambda = 2.5$ μm, (b) $\Lambda = 3$ μm, and (c) $\Lambda = 4$ μm [5.20].

$d_{\text{d}}/\Lambda = 0.9$. This fiber provides a Raman effective area which is only 1.5 μm$^2$ lower than that of the PCF with $d_{\text{d}} = 0$. A similar qualitative behavior of $A_{\text{eff}}^{\text{R}}$ and $\gamma_{\text{R}}$ has been found for different $\Lambda$ values, that is 3 and 4 μm, as shown in Fig. 5.10b and c, respectively. For example, the results obtained for the honeycomb PCFs with $\Lambda = 3$ μm, reported in Fig. 5.10b, suggest to choose $d_{\text{d}}/\Lambda = 1.2$ to double the $\gamma_{\text{R}}$ with respect to the silica core fiber, while just reducing $A_{\text{eff}}^{\text{R}}$ from 12.8 to 11.3 μm$^2$.

## 5.4 PCF Raman amplifiers

The performances of PCF-based Raman amplifiers have been studied with an accurate model which combines the calculation of the Raman gain coefficient

[5.26, 5.46] and the solution of the Raman propagation equations [5.4].

The Raman gain coefficient $\gamma_{\mathrm{R}}$ is related to the frequency separation $\Delta\nu$ between the interacting signals and it is calculated according to

$$\begin{aligned}\gamma_{\mathrm{R}}(\Delta\nu) &= \iint_S C_{\mathrm{SiSi}}(\Delta\nu)(1-2m(x,y))i_{\mathrm{s}}(x,y)i_{\mathrm{p}}(x,y)dxdy \\ &+ \iint_S C_{\mathrm{GeSi}}(\Delta\nu)2m(x,y)i_{\mathrm{s}}(x,y)i_{\mathrm{p}}(x,y)dxdy\,. \end{aligned} \tag{5.15}$$

This is a general definition of the Raman gain coefficient, since the dependency of the Raman spectra $C_{\mathrm{SiSi}}$ and $C_{\mathrm{GeSi}}$, and consequently of $\gamma_{\mathrm{R}}$, on the frequency separation $\Delta\nu$ is clearly expressed. As explained in Section 5.1, Eq. (5.11) derives directly from Eq. (5.15) when $\Delta\nu \simeq 13.2$ THz, which provides the Raman gain coefficient peak. It is important to underline that $C_{\mathrm{SiSi}}(\Delta\nu)$ and $C_{\mathrm{GeSi}}(\Delta\nu)$ are equal to 0 in the air-holes of the PCF cross-section, since the air contribution to the Raman amplification process is null.

The analysis of the Raman fiber amplifier performances is based on a set of propagation equations, which describe the forward and backward power evolutions along the fiber of pumps, signals, noise, and signal Rayleigh backscattering. The model includes the SRS and its amplification, the spontaneous Raman emission and its temperature dependence, the Rayleigh backscattering, the fiber loss, and the arbitrary interaction within pumps, signals and noise from either propagation directions [5.4, 5.5]. The accurate description of the Raman amplification of the WDM signals which simultaneously propagate along the PCF has been obtained by solving the propagation equations. Two methods, the Runge-Kutta algorithm and the Adams method, have been used to solve the differential equations, obtaining the same results.

### 5.4.1 Model for PCF Raman amplifiers

The Raman fiber amplifier is modeled through the following equations in the continuous-wave case:

$$\begin{aligned}\frac{dP^{\pm}(z,\lambda_i)}{dz} &= \pm\Big\{-\alpha(\lambda_i)+\sum_{j=1}^{N_T}\hat{\gamma}_{\mathrm{R}}(\lambda_i,\lambda_j)\big[P^{+}(z,\lambda_j) \\ &+P^{-}(z,\lambda_j)+n^{+}(z,\lambda_j)+n^{-}(z,\lambda_j)\big] \\ &+\sum_{j=i+1}^{N_T} 2K_{\mathrm{p}}E(\lambda_i,\lambda_j)\hat{\gamma}_{\mathrm{R}}(\lambda_i,\lambda_j)\Big\}P^{\pm}(z,\lambda_i)\,;\end{aligned} \tag{5.16}$$

$$\begin{aligned}\frac{dn^{\pm}(z,\lambda_i)}{dz} &= \pm\Big\{-\alpha(\lambda_i)+\sum_{j=1}^{N_T}\hat{\gamma}_{\mathrm{R}}(\lambda_i,\lambda_j)\big[P^{+}(z,\lambda_j)\\ &\quad +P^{-}(z,\lambda_j)+n^{+}(z,\lambda_j)+n^{-}(z,\lambda_j)\big]\\ &\quad +\sum_{j=i+1}^{N_T}2K_{\mathrm{p}}E(\lambda_i,\lambda_j)\hat{\gamma}_{\mathrm{R}}(\lambda_i,\lambda_j)\Big\}n^{\pm}(z,\lambda_i)\\ &\quad \pm\sum_{j=1}^{i-1}K_{\mathrm{p}}E(\lambda_i,\lambda_j)\hat{\gamma}_{\mathrm{R}}(\lambda_i,\lambda_j)\big[P^{+}(z,\lambda_j)\\ &\quad +P^{-}(z,\lambda_j)+n^{+}(z,\lambda_j)+n^{-}(z,\lambda_j)\big]\\ &\quad \pm r(\lambda_i)n^{\mp}(z,\lambda_i)\ ;\end{aligned} \tag{5.17}$$

$$\begin{aligned}\frac{dn^{\pm}_{\mathrm{SRB}}(z,\lambda_i)}{dz} &= \pm\Big\{-\alpha(\lambda_i)+\sum_{j=1}^{N_T}\hat{\gamma}_{\mathrm{R}}(\lambda_i,\lambda_j)\big[P^{+}(z,\lambda_j)\\ &\quad +P^{-}(z,\lambda_j)+n^{+}(z,\lambda_j)+n^{-}(z,\lambda_j)\big]\\ &\quad +\sum_{j=i+1}^{N_T}2K_{\mathrm{p}}E(\lambda_i,\lambda_j)\hat{\gamma}_{\mathrm{R}}(\lambda_i,\lambda_j)\Big\}n^{\pm}_{\mathrm{SRB}}(z,\lambda_i)\\ &\quad \pm r(\lambda_i)\big[n^{\mp}_{\mathrm{SRB}}(z,\lambda_i)+P^{\mp}(z,\lambda_i)\big]\ .\end{aligned} \tag{5.18}$$

$P^{\pm}(z,\lambda_i)$ is the forward/backward power of the pump or the signal at the wavelength $\lambda_i$ at the distance $z$ along the fiber. $n^{\pm}(z,\lambda_i)$ is the forward/backward power of the noise due to the amplified spontaneous Raman scattering and the Rayleigh backscattering. $n^{\pm}_{\mathrm{SRB}}(z,\lambda_i)$ is the forward/backward power of the Signal Rayleigh Backscattering (SRB). Notice that the contribution to the noise due to the Rayleigh backscattering of the signals has been distinguished from the other noise components, in order to evaluate the negative impact of the Double Rayleigh Backscattering (DRB) on the Raman fiber amplifier performances. In fact, some of the scattered light of the forward propagating signals is recaptured and it can pick up an additional power increase, due to the Raman amplification, while it is traveling backward along the fiber. After a rescattering and a recapture, the $n_{\mathrm{SRB}}$ becomes forward propagating, so it is effectively a DRB noise, and it can create the multiple path interference [5.1]. In the propagation equations $\hat{\gamma}_{\mathrm{R}}(\lambda_i,\lambda_j)$ is defined according to

$$\hat{\gamma}_{\mathrm{R}}(\lambda_i,\lambda_j)=\begin{cases}\dfrac{\gamma_{\mathrm{R}}(\Delta\nu_{i,j})}{K_{\mathrm{p}}}\cdot\dfrac{\lambda_{\mathrm{ref}}}{\lambda_j} & \lambda_i>\lambda_j\\ 0 & \lambda_i=\lambda_j\\ -\dfrac{\lambda_j}{\lambda_i}\cdot\dfrac{\gamma_{\mathrm{R}}(\Delta\nu_{i,j})}{K_{\mathrm{p}}}\cdot\dfrac{\lambda_{\mathrm{ref}}}{\lambda_i} & \lambda_i<\lambda_j\end{cases}, \quad (5.19)$$

where $\Delta\nu_{i,j}=|(c/\lambda_i)-(c/\lambda_j)|$, $K_{\mathrm{p}}$ is the polarization factor, here considered equal to 2, and $\lambda_{\mathrm{ref}}/\lambda_{p,\ p=i,j}$ opportunely scales the Raman gain coefficient when the pump wavelength is different from $\lambda_{\mathrm{ref}}$ [5.46]. The noise spectrum is 200 nm wide, from 1450 to 1650 nm, and it is divided into $N_{\mathrm{T}}=500$ slots of width $\Delta\lambda=0.4$ nm around the central wavelength $\lambda_i$. In Eqs. (5.16)–(5.18) the temperature dependence of the spontaneous Raman emission is described by

$$E(\lambda_i,\lambda_j)=\frac{hc^2\Delta\lambda}{\lambda_i(\lambda_i^2-\Delta\lambda^2/4)}\left[1+\frac{1}{\exp\left(\dfrac{hc|\lambda_i-\lambda_j|}{\lambda_i\lambda_j kT}\right)-1}\right], \quad (5.20)$$

where $h$ is the Planck's constant, $c$ is the light speed in vacuum, $k$ is the Boltzmann's constant, and $T$ is the absolute temperature of the fiber, fixed to 300.15 $K$. In Eqs. (5.16)–(5.18) $\alpha(\lambda_i)$ and $r(\lambda_i)$ are, respectively, the fiber background loss and the Rayleigh backscattering coefficient at the wavelength $\lambda_i$, both expressed in $m^{-1}$. $r(\lambda_i)$ is the product of the Rayleigh scattering loss $\alpha_s(\lambda_i)$ and the recapture fraction $B(\lambda_i)$ [5.1]. Due to the lack of experimental data for PCFs, the Rayleigh backscattering coefficient has been calculated by adapting to the PCF case an expression suited for single-mode optical fiber with arbitrary refractive index profiles and scattering-loss distributions [5.47]. The Rayleigh scattering loss is defined according to $\alpha_{\mathrm{s}}(\lambda_i)=C_{\mathrm{R}}/\lambda_i^4$, being $C_{\mathrm{R}}$ the Rayleigh scattering coefficient [5.25]. Notice that the value of $C_{\mathrm{R}}$ for pure silica glass, which is assumed to be 1 dB/km/µm$^4$ for a low-loss PCF [5.48], is modified by the presence of a $GeO_2$-doped area in the fiber cross-section, which causes a relative refractive index difference $\Delta n(x,y)$ [5.49]. By exploiting the accurate normalized intensity evaluated through the FEM solver, as reported in Appendix A, the Rayleigh backscattering coefficient for the PCFs here considered has been calculated according to

$$r(\lambda_i)=\alpha_{\mathrm{s}}(\lambda_i)B(\lambda_i)=\frac{3}{8\pi\lambda_i^2 n_{\mathrm{Si}}^2}\iint_S C_{\mathrm{R}}(1+44\Delta n(x,y))i^2(x,y)dxdy\ , \quad (5.21)$$

where $n_{\mathrm{Si}} = 1.45$ is the refractive index of silica. After solving the propagation equations Eqs. (5.16)–(5.18) through the Runge-Kutta algorithm or the Adams method, the net gain $G$ and the Noise Figure (NF) can be easily calculated [5.1]. The negative effect of the DRB on the signal at the wavelength $\lambda_i$ is evaluated by calculating the ratio between the SRB power $n^{+}_{\mathrm{SRB}}(\lambda_i)$ and the signal power $P^{+}(\lambda_i)$ at the PCF end, that is,

$$\mathrm{DRB}(\lambda_i) = 10 \cdot log_{10} \frac{n^{+}_{\mathrm{SRB}}(\lambda_i)}{P^{+}(\lambda_i)} . \tag{5.22}$$

### 5.4.2 Triangular PCF Raman amplifiers

After testing the proposed Raman fiber amplifier model with different kinds of standard single-mode fibers, a low-loss single-mode PCF with $\Lambda = 4.2$ µm and $d = 1.85$ µm, that is with $d/\Lambda = 0.44$, has been considered [5.50]. This fiber has been preferred to highly nonlinear PCFs with very small effective area, since, in spite of their high Raman gain coefficient, they prevent efficient Raman amplification, as reported in [5.12], because of the high attenuation values [5.51]. A $GeO_2$-doped area internally tangent to the first air-hole ring has been added in the PCF cross-section, in order to increase the Raman gain coefficient value [5.16]. The radius of the up-doped region has been fixed to 3 µm, while different germania concentrations have been taken into account, that is 6.3%, 8.7%, 11.2%, 15%, and 19.3%. The considered PCF background losses $\alpha(\lambda)$ have been experimentally measured in [5.50] and they have been assumed independent from the germania concetration. As reported in [5.50], they reach a minimum value of 0.58 dB/km at 1550 nm. In Fig. 5.11 the $\gamma_{\mathrm{R}}(\Delta\nu)$ values calculated with Eq. (5.15) for the all-silica PCF and the germania-doped ones are reported. Notice that the $\gamma_{\mathrm{R}}$ peak value increases with the $GeO_2$ concentration and slightly moves toward a lower value of the frequency shift $\Delta\nu$ between the interacting signals in the Raman amplification process. The maximum Raman coefficient value is 3.28 $(\mathrm{W{\cdot}km})^{-1}$ at $\Delta\nu/c = 433$ cm$^{-1}$ for the PCF with the highest germania concentration, 19.3%, and 1.33 $(\mathrm{W{\cdot}km})^{-1}$ at $\Delta\nu/c = 442$ cm$^{-1}$ for the all-silica PCF. As shown in Fig. 5.12, also the Rayleigh backscattering coefficient $r(\lambda)$ becomes higher when the PCF is more doped. When the $GeO_2$ concentration is 19.3%, $r$ is $5.3 \cdot 10^{-7}\mathrm{m}^{-1}$ at the shortest wavelength in the considered range, 1450 nm, and decreases to $4.1 \cdot 10^{-7}\mathrm{m}^{-1}$ at the longest one, that is 1650 nm.

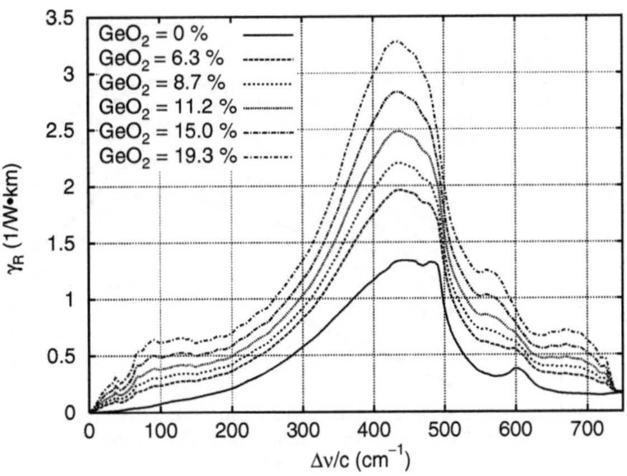


Figure **5.11**: Raman gain coefficient values for the PCFs with $\Lambda = 4.2$ μm, $d/\Lambda = 0.44$ and different germania concentrations in the doped area. The pump wavelength is 1455 nm [5.21].

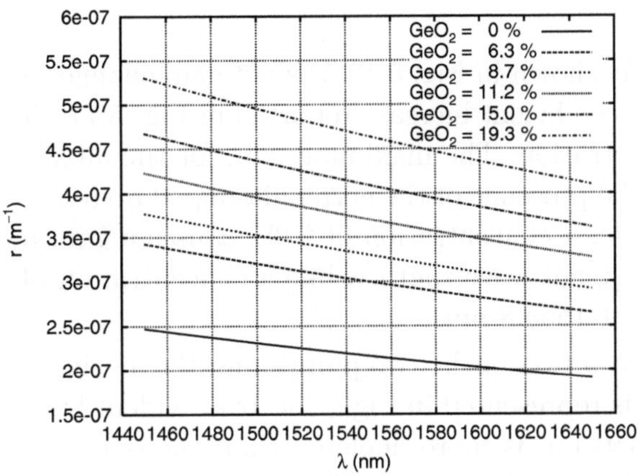


Figure **5.12**: Rayleigh backscattering coefficient versus the wavelength for the germania-doped PCFs with different concentrations [5.21].

In order to study the performances of the PCF Raman amplifier, 48 C-band channels with a frequency separation of about 100 GHz and an input power of −8 dBm/ch have been considered. Two counter-propagating pumps at 1450 and 1460 nm with a total input power of 933 mW have been used. The amplifier length has been chosen to be 6 km, since simulation results have shown that this is the optimum length for the PCF doped with the highest germania

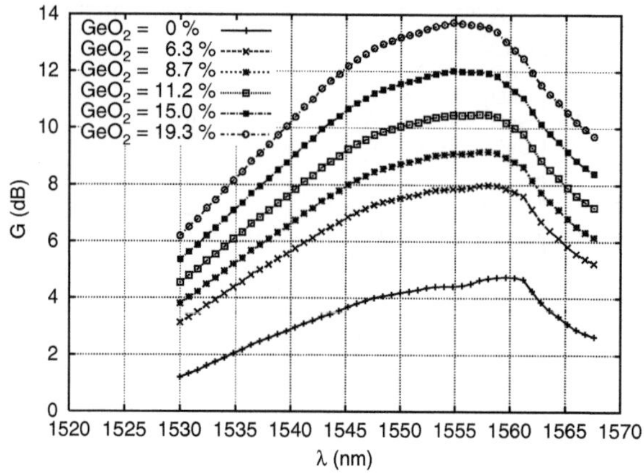


Figure **5.13**: Net gain $G$ versus the signal wavelength for the PCFs with $\Lambda =$ 4.2 μm, $d/\Lambda = 0.44$ and different germania concentrations in the doped area [5.21].

concentration. As shown in Fig. 5.13, the net gain changes as a consequence of the Raman gain coefficient increase reported in Fig. 5.11, due to the raising of the $GeO_2$ concentration. The maximum gain for the all-silica PCF is 4.74 dB at 1559.4 nm. The presence of the germania-doped area allows to increase this gain. For example, a concentration of 6.3% causes a gain increase of about 3 dB at the same signal wavelength, while a concentration of 19.3% provides a gain of 13.7 dB at 1554.8 nm.

The power evolution of the two pumps and the 48 signals which propagate along this PCF is represented in Fig. 5.14. It is evident from these results that the pump at 1450 nm is more depleted than the one at 1460 nm, since the lower wavelength pump has a strong Raman interaction with a higher number of the C band signals here considered.

The noise performances of the PCF Raman amplifier have been described by the NF and the DRB parameter. In Fig. 5.15 the DRB is shown as a function of the signal wavelength for the various $GeO_2$ doping levels here considered. For a fixed PCF geometry, the DRB parameter values increase with the germania concentration. For example, at 1550 nm the DRB is −57.6 dB for the all-silica PCF with $d/\Lambda = 0.44$ and $\Lambda = 4.2$ μm, and it increases to −40.5 dB when the germania concentration is 19.3%. Results have shown that the PCF amplifier with the best Raman gain, that is the one with the

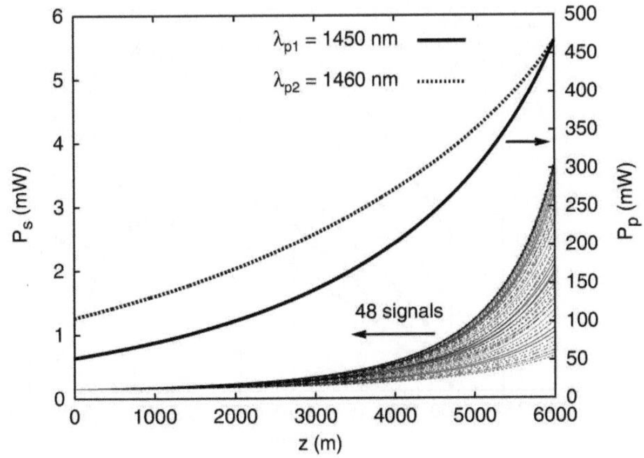


Figure **5.14**: Power evolution of the two counter-propagating pumps and the 48 C band signals along the doped PCF with a $GeO_2$ concentration of 19.3% [5.21].

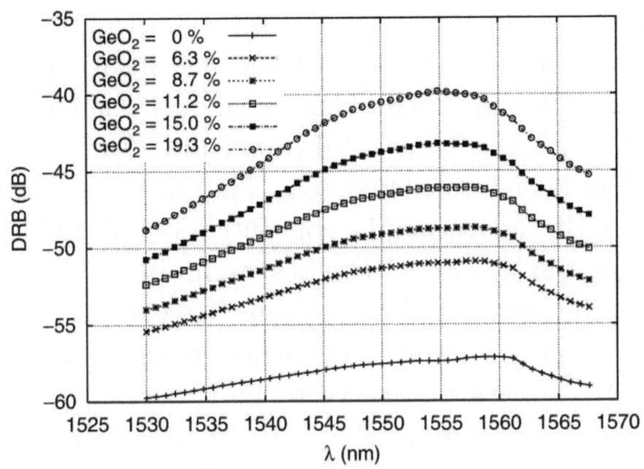


Figure **5.15**: DRB versus the signal wavelength for the PCF with $\Lambda = 4.2$ μm and different $GeO_2$ concentrations [5.21].

highest germania concentration, has also the worst NF values. In fact, the noise figure has a maximum value of 8.48 dB for the signal at 1530 nm, then it decreases as the wavelength increases, reaching the minimum value of 7.86 dB at 1567.6 nm.

The gain flexibility which can be obtained with Raman amplifiers has been analyzed by changing the wavelength separation $\Delta\lambda_p$ between the two

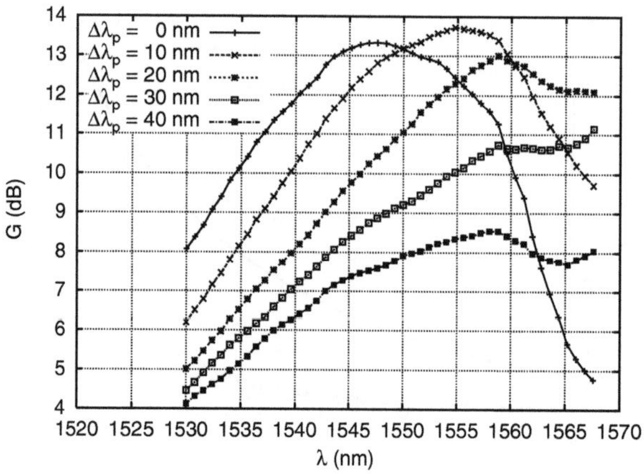


Figure **5.16**: Gain spectra for the doped PCF with a $GeO_2$ concentration of 19.3% for different $\Delta\lambda_p$ values [5.21].

counter-propagating pumps [5.2]. While the pump at the lower wavelength has been fixed to 1450 nm, the second one has been chosen at 1460, 1470, 1480, and 1490 nm. The pumping scheme with only one backward-propagating pump at 1450 nm with the same total power of 933 mW, which corresponds to $\Delta\lambda_p$=0 nm, has been considered too. The gain spectra obtained for the 6 km long PCF with the germania concentration of 19.3% are reported in Fig. 5.16 for different $\Delta\lambda_p$ values. The best pumping scheme, according to the net gain values, is the one with $\Delta\lambda_p$=10 nm. However, a gain of at least 13 dB has been reached for $\Delta\lambda_p \leq 20$ nm. The wavelength of the gain peak increases with the wavelength separation between the two pumps, being, respectively, 1547.6, 1554.8, and 1558.8 nm for $\Delta\lambda_p = 0$, 10, and 20 nm.

A final analysis of the PCF geometric parameter influence on the Raman amplifier performances has been carried out. A single-mode PCF with the same $d/\Lambda$ value, that is 0.44, and a smaller pitch $\Lambda = 3.2$ µm has been considered [5.50]. According to the measurements reported in [5.50], the PCF with the smaller pitch has a minimum loss of 1 dB/km at 1630 nm. For example, at 1550 nm, which is the wavelength of the central channel considered in the simulations, the background loss is 1.1 dB/km for the PCF with $\Lambda =$ 3.2 µm, almost twice the value for the PCF with the larger pitch. The Raman effective area $A_{\mathrm{eff}}^{\mathrm{R}}$ of this triangular PCF, that is 15.5 µm$^2$, is smaller than the one of the PCF with $\Lambda = 4.2$ µm, that is 25 µm$^2$. Since the Raman gain

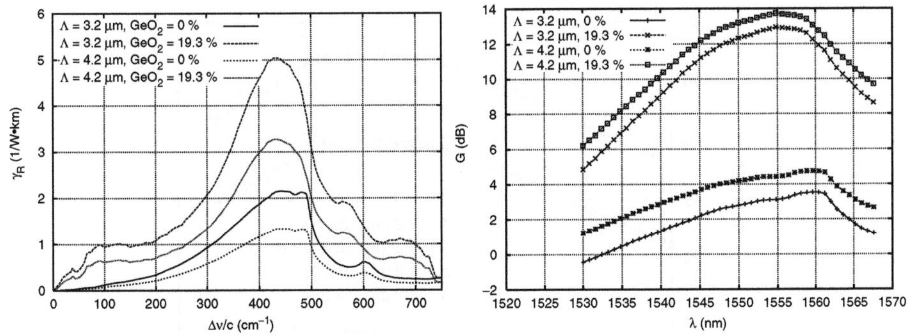


(a) (b)

Figure **5.17**: Comparison between (a) the Raman gain coefficient spectra and (b) the Raman gain spectra for the two PCFs with $d/\Lambda = 0.44$, without the germania-doped area and with the highest $GeO_2$ concentration [5.21].

coefficient is inversely related to the Raman effective area as previously stated in Section 5.2, the all-silica PCF with $\Lambda = 3.2$ μm has higher $\gamma_R$ values, as reported in Fig. 5.17a. The presence of the doped area, which is still internally tangent to the first air-hole ring, with a $GeO_2$ concentration of 19.3% causes higher $\gamma_R$ values in the triangular PCF with $\Lambda = 3.2$ μm. Figure 5.17a shows that a maximum Raman gain coefficient value of about 5 $(\text{W}\cdot\text{km})^{-1}$ has been obatined for the germania-doped PCF with the smallest pitch.

Finally, the Raman performances of the two triangular PCFs have been compared, by using the same amplifier configuration previously described, with $\Delta\lambda_p$=10 nm. Simulation results have shown that it is necessary to fix the length of the PCF with $\Lambda = 3.2$ μm to 5 km in order to maximize the gain obtained with a germania concentration of 19.3%. Notice that this optimum lenght is lower than the one calculated for the first PCF, which is 6 km. In Fig. 5.17b, the net gain spectra for the two all-silica PCFs and the two $GeO_2$-doped PCFs with the highest concentration are reported. It is interesting to notice that lower gain values have been obtained for the PCF with the higher $\gamma_R$, that is the one with $\Lambda = 3.2$ μm. The analysis performed has shown that the decrease of $G$ is caused by an increase of the background losses $\alpha(\lambda)$, due to the different geometric parameters. As reported in [5.50], the Rayleigh component of the background losses is higher when the pitch is reduced. In particular, the Rayleigh scattering coefficient $C_R$ for the PCF with $\Lambda = 3.2$ μm is almost twice the one of the PCF with $\Lambda = 4.2$ μm, causing

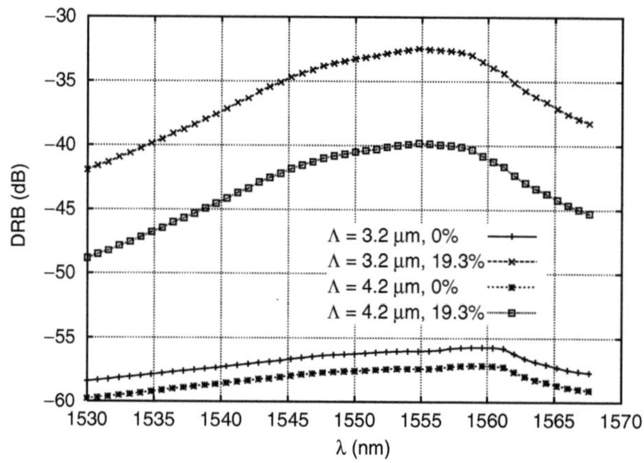


Figure **5.18**: Comparison between the DRB spectra for the two PCFs with $d/\Lambda = 0.44$, without the germania-doped area and with the highest $GeO_2$ concentration [5.21].

an increase of the Rayleigh backscattering coefficient $r(\lambda)$. As a consequence, the DRB values for the PCF with the smallest pitch are higher at all the signal wavelengths considered, as shown in Fig. 5.18. This is due also to the stronger field confinement, which can be obtained by reducing the pitch for a fixed $d/\Lambda$ value. Notice that the difference between the DRB values for the two all-silica PCFs is 1.33 dB at 1550 nm, and it becomes 7.26 dB if a 19.3% germania-doped region is added.

The PCF flexibility allows to considerably reduce the Raman effective area, and thus to increase the Raman gain. However, the background losses, related to the PCF geometry, can become a crucial factor for the amplifier performances. In order to clearly show this trade-off, the Raman performances of PCFs with different values of the Raman gain coefficient and of the background losses have been analyzed, taking into account a single signal at 1550 nm with an input power of −8 dBm and 933 mW of pump power at 1450 nm. The fiber length is fixed to 6 km and the background losses are those experimentally measured in [5.50] for the PCF with $d/\Lambda = 0.44$ and $\Lambda = 4.2$ μm. These loss values have been reduced or increased, multiplying by a factor equal to 0.5, 1, 2, or 5, in order to investigate how the fiber losses influence the amplifier design. Figure 5.19a shows the dependence of the Raman gain coefficient on the triangular PCF geometric parameters, that is $\Lambda$, which varies between 2.2

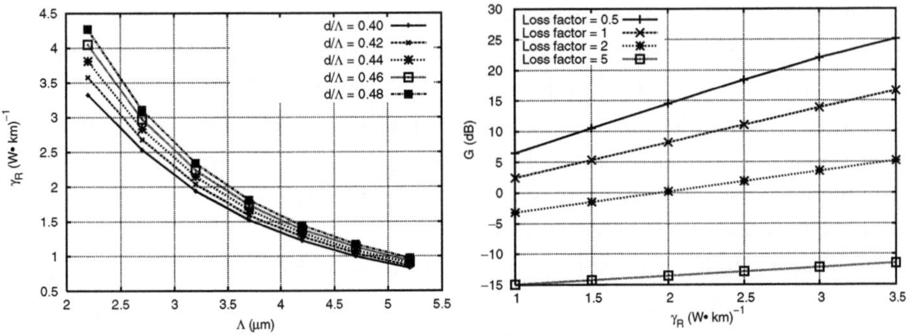


(a) (b)

Figure **5.19**: (a) Raman gain coefficient as a function of the PCF geometric parameters $d/\Lambda$ and $\Lambda$. (b) Design curve of the Raman gain at 1550 nm as a function of the Raman gain coefficient for different loss values [5.21].

and 5.2 μm, and $d/\Lambda$, chosen between 0.4 and 0.48. $\gamma_R$ values higher than that of the PCF with $d/\Lambda = 0.44$ and $\Lambda = 4.2$ μm can be obtained by decreasing the pitch or enlarging the air-hole diameter, as it has been already demonstrated in previous sections. The obtained $\gamma_R$ values have been used to compute the Raman gain for the various loss levels, as reported in Fig. 5.19b. Results have shown that good Raman gain values of at least 10 dB can be reached for $\gamma_R \geq 1.5$ (W·km)$^{-1}$, if the background losses are halved. On the contrary, an increase of the loss values dramatically reduces the Raman performances and high values of $\gamma_R$ are not enough to obtain gain for the fiber length considered. In conclusion, as long as the fabrication process will not provide a further background loss reduction, a trade-off between the losses and the PCF effective area design has to be found.

## 5.5 Impact of background losses on PCF Raman amplifiers

As demonstrated by the results reported in Fig. 5.19, despite their high Raman gain coefficient values, the possibility to successfully use nonlinear PCFs as Raman amplifiers for telecommunication applications is currently limited by their high attenuation values [5.12].

Thus, it is interesting to analyze the influence of the background losses on the gain and noise performances of triangular PCF Raman amplifiers, by considering the three low-loss fibers presented in [5.52]. These PCFs are particularly suitable for this kind of analysis, since they have almost the same geometric parameters and differ only for the background and the OH-absorption losses. The present study has been carried out through the model previously described in Section 5.4.1.

The first PCF, referred to as fiber A in [5.52], has $d/\Lambda = 0.625$ and $\Lambda = 4$ μm. Fibers B, and C have the same pitch value and a slightly lower air-filling fraction, that is $d/\Lambda = 0.6$. The background loss values of PCFs A, B, and C, which have been measured in [5.52], are compared in Fig. 5.20a for the wavelength range 1450–1650 nm. Starting from the Rayleigh scattering coefficients reported in [5.52], that is 1.0, 2.3, and 1.9 dB/km/μm$^4$ for PCF A, B, and C, respectively, the Rayleigh backscattering coefficient spectra have been computed, as shown in Fig. 5.20b. The evaluation of the Raman gain coefficient has provided peak values of about 2.06 $(\text{W}\cdot\text{km})^{-1}$ for fiber A and 1.97 $(\text{W}\cdot\text{km})^{-1}$ for fibers B and C. In the studied PCF Raman amplifiers a 1 W counterpropagating pump at 1450 nm and 40 channels between 1540.4 and 1571.6 nm, with a frequency separation of about 100 GHz and an input power of −20 dBm/ch, have been considered.

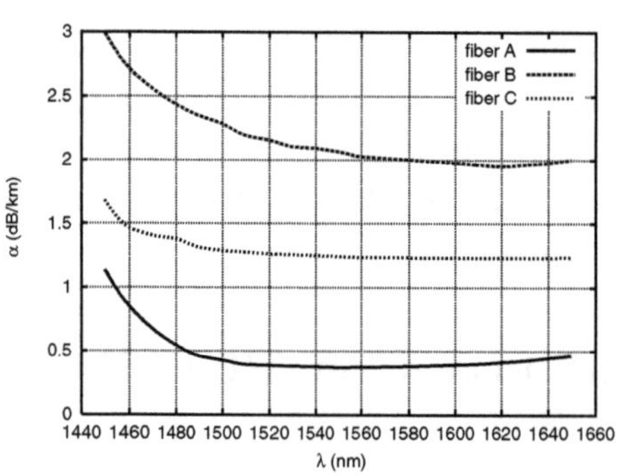


(a)

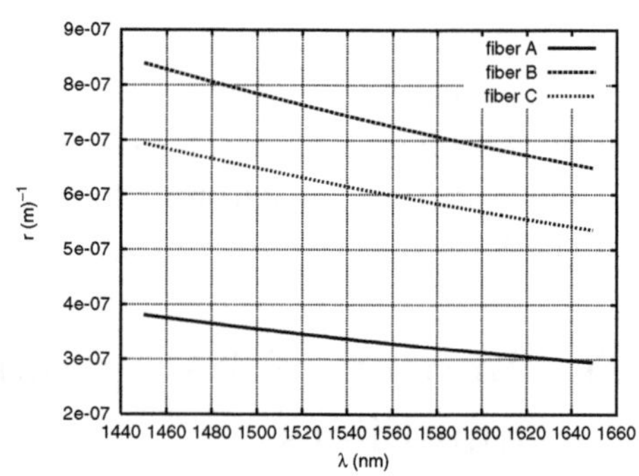


(b)

Figure **5.20**: (a) Background losses and (b) Rayleigh backscattering coefficient spectra for the three considered PCFs [5.22].

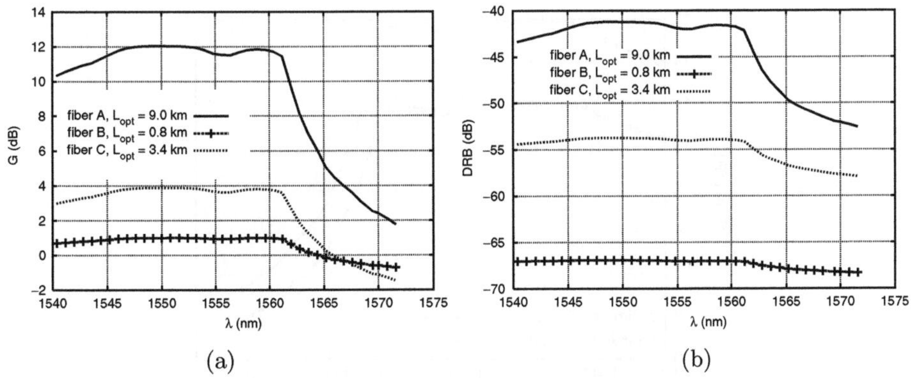


Figure **5.21**: (a) Raman gain spectra and (b) DRB for the three PCFs with the optimum length [5.22].

As shown in Fig. 5.21a, the maximum Raman gain, 12 dB, has been achieved with the 9 km long fiber A, which has the lowest losses, in particular 0.37 dB/km at 1550 nm. Unfortunately, as a consequence of its good gain performances, fiber A has also the highest DRB values, reported in Fig. 5.21b, even if its Rayleigh backscattering coefficient is the lowest for all the considered wavelengths, as shown in Fig. 5.20b. Notice that an increase of the losses causes a significant decrease of the maximum Raman gain. In fact, the best gain for fiber B, the one with the highest losses, is only 1 dB, obtained for a 0.8 km fiber length. Fiber C, as expected, presents intermediate values for both the gain and the DRB.

Once optimized the background losses, a further improvement can be achieved with a reduction of the OH-absorption losses [5.52]. For example, a fiber with the loss spectra of fiber C, which has been fabricated in order to reduce the OH-absorption peak around 1380 nm, down scaled to the fiber A loss level has been considered. Simulation results, assuming the same PCF A $\gamma_R(\lambda)$ and $r(\lambda)$, have demonstrated that a loss reduction of only 0.33 dB/km at the pump wavelength causes much better Raman performances with respect to PCF A, as shown in Fig. 5.22. In fact, a gain increase of about 4 dB can be reached when the optimized fiber length is 9 km. Due to the higher pump efficiency, the same maximum gain previously reached with a 9 km long fiber A, that is 12 dB, can be obtained with a half-length optimized PCF, that is 4.5 km long.

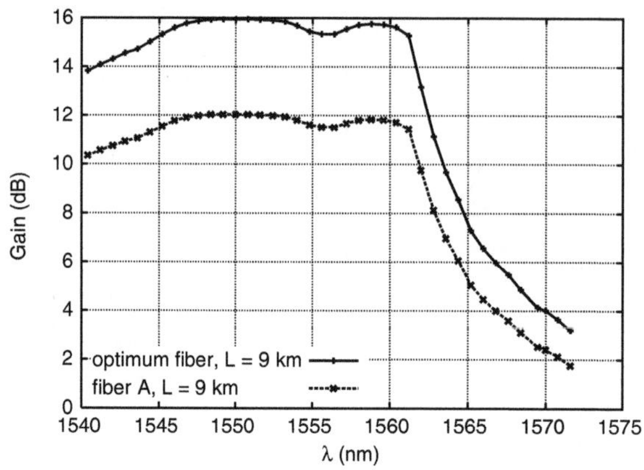


Figure **5.22**: Raman gain spectra obtained with fiber A and with the optimized PCF [5.22].

## 5.6 Multipump PCF Raman amplifiers

In the analysis here presented a broadband approach has been applied to one of the ultralow-loss triangular PCF considered in Section 5.5, that is fiber A [5.52], in order to provide, for the first time, a preliminary investigation of the performances of PCF-based multipump Raman amplifiers. The attention is focused on multipump schemes, as the spectral flexibility of Raman amplification allows to obtain broadband amplification by combining multiple pump wavelengths. In particular, by using the superposition rule proposed in [5.5] the wavelengths and the power levels of the Raman pumps can be optimized and, for example, amplifiers with gain bandwidths greater than 100 nm have been already demonstrated by using conventional fibers [5.3].

The triangular PCF here considered to study the Raman amplifier performances is the one with $\Lambda = 4$ μm and $d/\Lambda = 0.625$ [5.52], described in Section 5.5 as fiber A. It is important to recall that this fiber has a maximum Raman gain coefficient of 2.06 $(\mathrm{W \cdot km})^{-1}$ and it has been chosen since it is an ultralow-loss PCF, whose background loss spectrum is reported in Fig. 5.20a. The PCF-based Raman amplifier analyzed is characterized by a 40-channel input WDM spectrum extended between 1540.4 and 1571.6 nm, with a frequency separation of about 100 GHz and an input power of −20 dBm per channel. As in Section 5.5 the fiber length has been chosen equal to 9 km, lower than the real drawn fiber length of 10 km [5.52]. The performances of

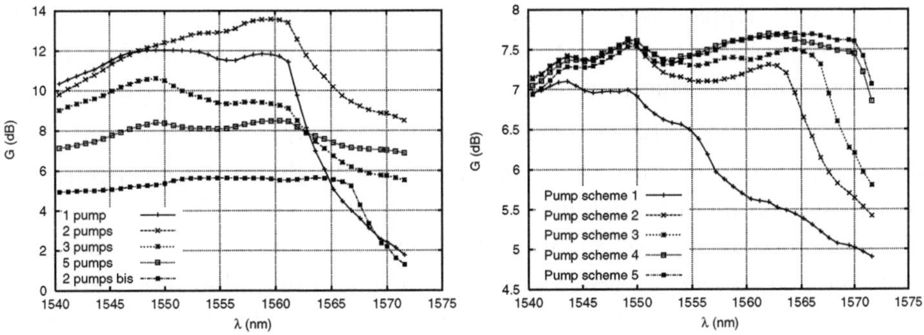


(a) (b)

Figure **5.23**: (a) Raman gain for 40 input signals, employing one, two, three, and five backward pumps. (b) Raman gain spectra for six pumps with equal power of 167 mW each. $\lambda_1$ is fixed at 1430 nm, $\lambda_2$ at 1435 nm, $\lambda_3$ at 1440 nm and $\lambda_5$ at 1461 nm. For schemes 1–4, $\lambda_6 = 1464$ nm while $\lambda_4$ is 1445, 1452, 1455 and 1458 nm, respectively. For scheme 5, $\lambda_4 = 1458\ nm$ and $\lambda_6 = 1466$nm [5.24].

the triangular PCF Raman amplifiers have been investigated by considering different pumping configurations, that is by changing the number of pumps, their wavelength and the power associated with each one.

Firstly, schemes with two, three, and five pumps have been considered, with a constant total pumping power of 1 W and a constant wavelength spacing of 10 nm. The simulated schemes present two pumps at 1450 and 1460 nm, three pumps at 1440, 1450, and 1460 nm, and five pumps at 1430, 1440, 1450, 1460 and 1470 nm. The calculated gain spectra are shown in Fig. 5.23a, together with the gain curve obtained for a single pump at 1450 nm for comparison purposes. Notice that, by adding the second pump at 1460 nm, the gain increases for the wavelengths above 1550 nm, reaching a maximum value of 13.6 dB at 1560 nm. Increasing the number of pumps lowers the peak value, but flattens the gain spectrum, reaching a maximum value of 8.5 dB at 1560.4 nm, with a minimum of 6.9 dB at 1571.6 nm in the five-pump scheme. Although this is the best case so far, the gain ripple results about 1.6 dB in the considered signal wavelength range. Moreover, it is interesting to compare these results with those providing the optimum flatness over the C band when using two pumps, that is 1428 and 1455 nm [5.5]. The curve, obtained, respectively, with 548 and 452 mW pump powers and labeled as *2 pumps bis* in Fig. 5.23a, is very flat, with ripples lower than 0.5 dB, but the maximum gain is just around 5.5 dB. By properly changing the power levels, it can be increased up to 7 dB, to the detriment of the flatness profile, whose ripple exceeds 1 dB.

In order to achieve better performances, still keeping fixed the total amount of power launched into the PCF, an amplifier with six pumps has been considered, with a number of different wavelength configurations. The wavelengths of the first pumps, that is $\lambda_1$, $\lambda_2$, $\lambda_3$, and $\lambda_5$, have been fixed, while $\lambda_4$ and $\lambda_6$ have been changed in thc ranges 1445–1458 nm, and 1464–1466 nm, respectively. The total optical power of 1 W has been evenly divided among the six pumps. As shown by the spectra of the first four schemes reported in Fig. 5.23b, by progressively upshifting the wavelength $\lambda_4$, while fixing $\lambda_6$ to 1464 nm, the gain spectrum is raised only in the high wavelength range. In fact, while in scheme 1 the gain difference over the full range is almost 2 dB, the gain ripple lowers to about 0.5 dB in the reduced ranges 1540–1564 nm and 1540–1567 nm for schemes 2 and 3, respectively. By further raising $\lambda_4$ in scheme 4, a gain ripple of 0.84 dB has been obtained for all the signal wavelengths considered. Moreover, by choosing a higher value of 1466 nm for $\lambda_6$ in scheme 5, the gain performance is only slightly influenced, showing a peak of 7.7 dB around 1564 nm, while the flatness improves to 0.76 dB.

Since the pump interaction with the signals depends on the pump spectral separation, in particular since pumps with longer wavelengths mainly interact with longer wavelength signals, unequal power allocation between pumps can improve the PCF-based Raman amplifier gain flatness. On the other hand, signals with shorter wavelengths typically receive gain contributions more uniformly from all the pumps. Several simulations have been thus performed by distributing the power unevenly among the pumps, while keeping total power equal to 1 W, in order to obtain a gain spectrum as flat as possible with the wavelength distributions of schemes 1 and 3. The best gain spectra obtained are shown in Fig. 5.24 a and b and the corresponding power distributions are reported in Table 5.3. Notice that high pump power should be supplied at wavelengths near or higher than 1460 nm, that is at wavelengths $\lambda_5$, and $\lambda_6$, in order to increase the gain at higher signal wavelengths with respect to the even power distribution. Moreover, pump 1 should also be allocated a high power share, since it is the pump wavelength that suffers highest background losses, as shown in Fig. 5.20a. As a consequence, in the considered cases $\lambda_1$, $\lambda_5$, and $\lambda_6$ almost always account for more than two thirds of the total pump power. One important criterion in the design of PCFs for Raman amplification is thus

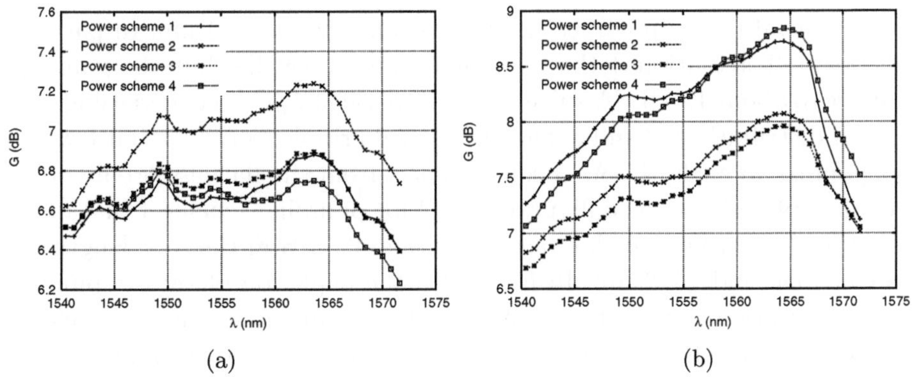


Figure **5.24**: Gain spectra for pumping scheme (a) 1 and (b) 3 with varying pump powers, as reported in Table 5.3 [5.24].

Table 5.3: Pump powers for selected pumping schemes [5.24].

| | $P_{p1}$ (mW) | $P_{p2}$ (mW) | $P_{p3}$ (mW) | $P_{p4}$ (mW) | $P_{p5}$ (mW) | $P_{p6}$ (mW) |
|---|---|---|---|---|---|---|
| **Pumping scheme 1** | | | | | | |
| **1** | 250 | 150 | 120 | 50 | 210 | 220 |
| z‘**2** | 240 | 130 | 120 | 60 | 230 | 220 |
| **3** | 250 | 140 | 120 | 60 | 220 | 210 |
| **4** | 250 | 140 | 120 | 70 | 210 | 210 |
| **Pumping scheme 3** | | | | | | |
| **1** | 180 | 110 | 140 | 150 | 220 | 200 |
| **2** | 220 | 130 | 130 | 80 | 220 | 220 |
| **3** | 240 | 130 | 120 | 60 | 230 | 220 |
| **4** | 200 | 110 | 120 | 120 | 240 | 210 |

the reduction of the OH peak, which considerably reduces the low-wavelength pump efficiency.

All the spectra belonging to scheme 1, shown in Fig. 5.24a, have peaks at 1549 and 1562 nm, with a maximum gain of 7.2 dB for the latter wavelength using power distribution 2. However, for all the considered configurations the gain ripple is between 0.5 and 0.65 dB. The effect of upshifting the wavelength $\lambda_4$ is apparent in the spectra belonging to scheme 3, shown in Fig. 5.24b.

Notice that all the gain curves exhibit a peak gain at 1564 nm. In particular, with power distribution 4 the maximum gain is 8.8 dB, but the gain ripple increases to about 1.8 dB. A substantial decrease of $P_{p4}$ while increasing $P_{p1}$, as in power distributions 2 and 3, flattens the gain ripple to 1.3 dB, but lowers the peak gain to about 8 dB. As a consequence, from the flatness profile point of view, the optimal pump configuration is 3 for pumping scheme 1 in Table 5.3, being the gain flatness in this case lower than 0.5 dB in the wavelength range between 1540 and 1572 nm. In Fig. 5.25a the depletion of the six pumps for this configuration is reported. Notice that the pumps at longer wavelengths are lower absorbed, because of the weaker losses due to OH ions and the interaction with a minor number of signals.

Finally, notice that the 40 channels so far considered cover the C band and a small part of the L band. If only channels located in L band, from 1590 to 1622 nm, are considered, a much higher gain is obtainable. Simulation results reported in Fig. 5.25b have shown that a mean gain higher than 13.5 dB with a ripple lower than 0.6 dB can be obtained in L band with six pumps at 1475, 1482, 1490, 1505, 1510, and 1520 nm, with power allocations of 210, 150, 100, 150, 200, and 190 mW, respectively. For comparison, in the same figure also the gain obtained in the C band with the best pumping scheme, that is pump configuration 3 for pumping scheme 1, is reported. It is interesting

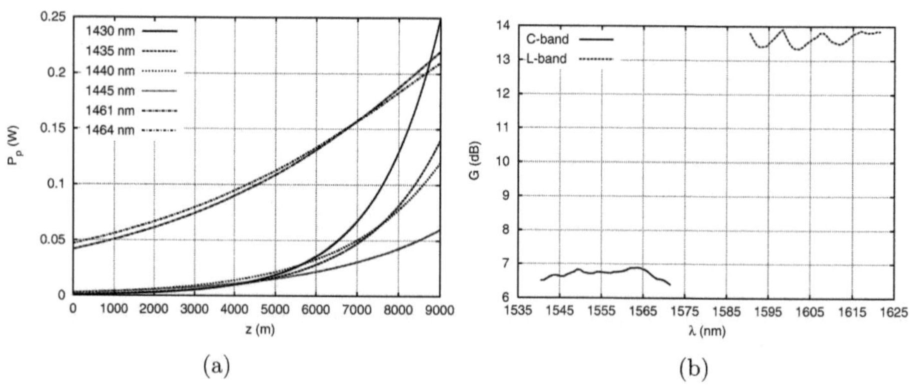


Figure **5.25**: (a) Depletion of the six backward pumps for the PCF-based Raman amplifier with the best pumping scheme, that is with pump configuration 3 for pumping scheme 1, and (b) flat Raman gain spectra in C and L bands [5.24].

to underline that in L band a much higher gain is obtained with the same total power, because the losses at the higher pumps wavelengths are more than halved in comparison with the losses near 1430 nm, where the first pump wavelength for the C band is located.

As a final remark, it must be observed that Raman amplifiers based on PCFs are still quite far from commercial exploitation due to actual constrains which limit their performances, as observed in the previous discussions. In particular, the background attenuation of the fiber, especially at the pump wavelength, the required total pump budget, the complex fiber design, and the final achievable gain, result in PCF Raman amplifiers not competitive with those based on standard technology fibers.

# Bibliography

[5.1] J. Bromage, "Raman amplification for fiber communication systems," in *Proc. Optical Fiber Communications Conference OFC 2003*, Atlanta, Georgia, USA, Mar. 23–28, 2003, pp. TuC1–1–TuC1–25.

[5.2] A. Evans, "Applications of Raman gain in optical transmission systems," in *Proc. European Conference on Optical Communication ECOC 2003*, Rimini, Italy, Sept. 21–25, 2003, paper Tutorial Mo3.3, pp. 156–183.

[5.3] Y. Emori, K. Tanaka, and S. Namiki, "100 *nm* bandwidth flat-gain Raman amplifiers pumped and gain-equalized by 12-wavelength-channel WDM laser diode unit," *Electronics Letters*, vol. 35, pp. 1355–1356, Aug. 1999.

[5.4] H. D. Kidorf, K. Rottwitt, M. Nissov, M. Ma, and E. Rabarijaona, "Pump interactions in a 100-*nm* bandwidth Raman amplifier," *IEEE Photonics Technology Letters*, vol. 11, pp. 530–532, May 1999.

[5.5] S. Namiki and Y. Emori, "Ultrabroad-band raman amplifiers pumped and gain-equalized by wavelength-division-multiplexed high-power laser diodes," *IEEE Journal of Selected Topics in Quantum Electronics*, vol. 7, pp. 3–16, Jan./Feb. 2001.

[5.6] P. B. Hansen, L. Eskildsen, S. G. Grubb, A. J. Stentz, T. A. Strasser, J. Judkins, J. J. Demarco, R. Pedrazzani, and D. J. Digiovanni, "Capacity upgrades of transmission systems by Raman amplification," *IEEE Photonics Technology Letters*, vol. 9, pp. 262–264, Feb. 1997.

[5.7] C. Fludger, A. Maroney, N. Jolley, and R. Mears, "An analysis of the improvements in OSNR from distributed Raman amplifiers using modern transmission fibers," in *Proc. Optical Fiber Communications Conference OFC 2000*, Mar. 7–10, 2000, paper FF2.

[5.8] K. P. Hansen and R. E. Kristiansen, "Supercontinuum Generation in Photonic Crystal Fibers," Crystal Fibre A/S, Tech. Rep., 2005.

[5.9] R. E. Kristiansen, K. P. Hansen, J. Broeng, P. M. W. Skovgaard, M. D. Nielsen, A. Petersson, T. P. Hansen, B. Mangan, C. Jakobsen, and H. R. Simonsen, "Microstructured fibers and their applications," in *Proc. Reuniòn Española de Optoelectrònica OPTOEL 2005*, Elche, Spain, July 13–15, 2005.

[5.10] A. Ortigosa-Blanch, J. C. Knight, and P. St. J. Russell, "Pulse breaking and supercontinuum generation with 200-$fs$ pump pulses in photonic crystal fibers," *Journal of Optical Society of America B*, vol. 19, pp. 2567–2572, Nov. 2002.

[5.11] J. C. Knight, "Dispersion and nonlinearity in photonic crystal fibres," in *Proc. Summer-School on Advanced Glass-Based Nano-Photonics POWAG 2004*, Bath, UK, July 12–16, 2004.

[5.12] C. J. S. de Matos, K. P. Hansen, and J. R. Taylor, "Experimental characterisation of Raman gain efficiency of holey fibre," *Electronics Letters*, vol. 39, pp. 424–425, Mar. 2003.

[5.13] J. Nilsson, R. Selvas, W. Belardi, J. H. Lee, Z. Yusoff, T. M. Monro, and D. J. Richardson, "Continous-wave pumped holey fiber Raman laser," in *Proc. Optical Fiber Communications Conference OFC 2002*, Anaheim, California, USA, Mar. 17–22, 2002, paper WR6, pp. 315–317.

[5.14] Z. Yusoff, J. H. Lee, W. Belardi, T. M. Monro, P. C. Teh, and D. J. Richardson, "Raman effects in a highly nonlinear holey fiber: amplification and modulation," *Optics Letters*, vol. 27, pp. 424–426, Mar. 2002.

[5.15] J. Hewett, "Fibre cuts Raman threshold," *Opto and Laser Europe*, vol. 100, p. 14, Nov. 2002.

[5.16] M. Fuochi, F. Poli, S. Selleri, A. Cucinotta, and L. Vincetti, "Study of Raman amplification properties in triangular photonic crystal fibers," *IEEE/OSA Journal of Lightwave Technology*, vol. 21, pp. 2247–2254, July 2003.

[5.17] M. Fuochi, F. Poli, S. Selleri, and A. Cucinotta, "Photonic crystal fibers for raman amplification," in *Proc. Progress in Electromagnetics Research Symposium PIERS 2003*, Honolulu, Hawaii, USA, Oct. 13–16, 2003.

[5.18] M. Fuochi, F. Poli, S. Selleri, A. Cucinotta, and L. Vincetti, "Raman amplification properties of silica and tellurite photonic crystal fibers," in *Proc. European Conference on Optical Communication ECOC 2003*, Rimini, Italy, Sept. 21–25, 2003.

[5.19] M. Bottacini, F. Poli, A. Cucinotta, S. Selleri, and A. H. Bouk, "Effective area tailoring in triangular photonic crystal fibers," in *Proc. Progress in Electromagnetics Research Symposium PIERS 2004*, Pisa, Italy, Mar. 28–31, 2004.

[5.20] S. Selleri, F. Poli, and A. Cucinotta, "Raman gain coefficient of solid-core honeycomb photonic crystal fibers," in *Proc. Laser and Ectro-Optics Society Annual Meeting LEOS 2004*, Puerto Rico, USA, Nov. 7–11, 2004.

[5.21] M. Bottacini, F. Poli, A. Cucinotta, and S. Selleri, "Modeling of photonic crystal fiber Raman amplifiers," *IEEE/OSA Journal of Lightwave Technology*, vol. 22, pp. 1707–1713, July 2004.

[5.22] M. Bottacini, S. Selleri, F. Poli, A. Cucinotta, and M. Foroni, "Impact of background losses on photonic crystal fiber Raman amplifier," in *Proc. Laser and Ectro-Optics Society Annual Meeting LEOS 2004*, Puerto Rico, USA, Nov. 7–11, 2004.

[5.23] S. Selleri, A. Cucinotta, M. Bottacini, F. Poli, and M. Foroni, "Gain flatness in photonic crystal fiber Raman amplifier," in *Proc. International Congress on Optics and Optoelectronics SPIE-COO 2005*, Warsaw, Poland, Aug. 28–Sept. 2, 2005.

[5.24] F. Poli, L. Rosa, M. Bottacini, M. Foroni, A. Cucinotta, and S. Selleri, "Multi-pump flattened-gain Raman amplifiers based on photonic-crystal fibers," *IEEE Photonics Technology Letters*, vol. 17, pp. 2556–2558, Dec. 2005.

[5.25] G. P. Agrawal, *Nonlinear Fiber Optics.* New York: Academic, 2001.

[5.26] J. Bromage, K. Rottwitt, and M. E. Lines, "A method to predict the Raman gain spectra of germanosilicate fibers with arbitrary index profile," *IEEE Photonics Technology Letters*, vol. 14, pp. 24–26, Jan. 2002.

[5.27] N. A. Mortensen, "Effective area of photonic crystal fiber," *Optics Express*, vol. 10, pp. 341–348, Apr. 2002. Available at: http://www.opticsexpress.org/abstract.cfm?URI=OPEX-10-7-341

[5.28] P. Petropoulos, T. M. Monro, W. Belardi, K. Furusawa, J. H. Lee, and D. J. Richardson, "2R-regenerative all-optical switch based on a highly nonlinear holey fiber," *Optics Letters*, vol. 26, pp. 1233–1235, Aug. 2001.

[5.29] J. H. Lee, Z. Yusoff, W. Belardi, M. Ibsen, T. M. Monro, and D. J. Richardson, "Investigation of Brillouin effects in small-core holey optical fiber: lasing and scattering," *Optics Letters*, vol. 27, pp. 927–929, June 2002.

[5.30] F. L. Galeener, J. C. Mikkelsen, R. H. Geils, and W. H. Mosby, "The relative Raman cross sections of vitreous $SiO_2$, $GeO_2$, $B_2O_3$ and $P_2O_5$," *Applied Physics Letters*, vol. 32, pp. 34–36, Jan. 1978.

[5.31] B. J. Ainslie, S. T. Davey, W. J. M. Rothwell, B. Wakefield, and D. L. Williams, "Optical gain spectrum of $GeO_2 - SiO_2$ Raman fibre amplifiers," *IEE Proceedings Optoelectronics*, vol. 136, pp. 301–306, Dec. 1989.

[5.32] *Highly Non-Linear Fiber for Discrete Raman Amplifier*, Sumitomo Electric Lightwave Corp.

[5.33] J. C. Knight, J. Arriaga, T. A. Birks, A. Ortigosa-Blanch, W. J. Wadsworth, and P. St. J. Russell, "Anomalous dispersion in photonic crystal fiber," *IEEE Photonics Technology Letters*, vol. 12, pp. 807–809, July 2000.

[5.34] K. Kato, H. Masuda, A. Mori, K. Oikawa, K. Shikano, and M. Shimizu, "Ultra-wideband tellurite-based Raman fibre amplifier," *Electronics Letters*, vol. 37, pp. 1442–1443, Nov. 2001.

[5.35] A. Mori, M. Shimizu, and H. Masuda, "Ultra-wideband tellurite-based fiber Raman amplifiers," in *Proc. Optical Fiber Communications Conference OFC 2003*, Atlanta, Georgia, USA, Mar. 23–28, 2003, pp. 427–429.

[5.36] E. S. Hu, Y.-L. Hsueh, M. E. Marhic, and L. G. Kazovsky, "Design of highly-nonlinear tellurite fibers with zero dispersion near 1550 nm," in *Proc. European Conference on Optical Communication ECOC 2002*, Copenhagen, Denmark, Sept. 8–12, 2002, paper 3.2.3.

[5.37] V. V. R. K. Kumar, A. K. George, J. C. Knight, and P. St. J. Russell, "Tellurite photonic crystal fiber," *Optics Express*, vol. 11, pp. 2641–2645, Oct. 2003. Available at: http://www.opticsexpress.org/abstract.cfm?URI=OPEX-11-20-2641

[5.38] D. Ferrarini, L. Vincetti, M. Zoboli, A. Cucinotta, and S. Selleri, "Leakage properties of photonic crystal fibers," *Optics Express*, vol. 10, pp. 1314–1319, Nov. 2002. Available at: http://www.opticsexpress.org/abstract.cfm?URI=OPEX-10-23-1314

[5.39] J. C. Knight, J. Broeng, T. A. Birks, and P. St. J. Russell, "Photonic band gap guidance in optical fibers," *Science*, vol. 282, pp. 1476–1478, Nov. 1998.

[5.40] A. Cucinotta, F. Poli, S. Selleri, L. Vincetti, and M. Zoboli, "Amplification properties of $Er^{3+}$-doped photonic crystal fibers," *IEEE/OSA Journal of Lightwave Technology*, vol. 21, pp. 782–788, Mar. 2003.

[5.41] J. Lægsgaard, N. A. Mortensen, and A. Bjarklev, "Mode areas and field-energy distribution in honeycomb photonic bandgap fibers," *Journal of Optical Society of America B*, vol. 20, pp. 2037–2045, Oct. 2003.

[5.42] J. Lægsgaard and A. Bjarklev, "Doped photonic bandgap fibers for short-wavelength nonlinear devices," *Optics Letters*, vol. 28, pp. 783–785, May 2003.

[5.43] T. P. Hansen, J. Broeng, and A. Bjarklev, "Solid-core photonic bandgap fiber with large anomalous dispersion," in *Proc. Optical Fiber Communications Conference OFC 2003*, Atlanta, Georgia, USA, Mar. 23–28, 2003, pp. 700–701.

[5.44] N. A. Mortensen, M. D. Nielsen, J. R. Folkenberg, C. Jakobsen, and H. R. Simonsen, "Photonic crystal fiber with a hybrid honeycomb cladding," *Optics Express*, vol. 12, pp. 468–472, Feb. 2004. Available at: http://www.opticsexpress.org/abstract.cfm?URI=OPEX-12-3-468

[5.45] S. G. Johnson and J. D. Joannopoulos, "Block-iterative frequency-domain methods for Maxwell's equations in a planewave basis," *Optics Express*, vol. 8, pp. 173–179, Jan. 2001. Available at: http://www.opticsexpress.org/abstract.cfm?URI=OPEX-8-3-173

[5.46] K. Rottwitt, J. Bromage, A. J. Stentz, L. Leng, M. E. Lines, and H. Smith, "Scaling of the Raman gain coefficient: applications to germanosilicate fibers," *IEEE/OSA Journal of Lightwave Technology*, vol. 21, pp. 1652–1662, July 2003.

[5.47] A. H. Hartog and M. P. Gold, "On the theory of backscattering in single-mode optical fibers," *IEEE/OSA Journal of Lightwave Technology*, vol. 2, pp. 76–82, Apr. 1984.

[5.48] K. Tajima, J. Zhou, K. Kurokawa, and K. Nakajima, "Low water peak photonic crystal fibres," in *Proc. European Conference on Optical Communication ECOC 2003*, Rimini, Italy, Sept. 21–25, 2003, paper Th4.1.6.

[5.49] W. Zhi, R. Guobin, L. Shuqin, and J. Shuisheng, "Loss properties due to Rayleigh scattering in different types of fiber," *Optics Express*, vol. 11, pp. 39–47, Jan. 2003. Available at: http://www.opticsexpress.org/abstract.cfm?URI=OPEX-11-1-39

[5.50] L. Farr, J. C. Knight, B. J. Mangan, and P. J. Roberts, "Low loss photonic crystal fibre," in *Proc. European Conference on Optical Communication ECOC 2002*, Copenhagen, Denmark, Sept. 8–12, 2002, paper PD1.3.

[5.51] Nonlinear photonic crystal fibers – Crystal Fibre A/S. Available at: http://www.crystal-fibre.com/products/nonlinear.shtm

[5.52] K. Tajima, J. Zhou, K. Nakajima, and K. Sato, "Ultralow loss and long length photonic crystal fiber," *IEEE/OSA Journal of Lightwave Technology*, vol. 22, pp. 7–10, Jan. 2004.

# Chapter 6

# Erbium-doped fiber amplifiers

In recent years, PCFs have emerged as an attractive alternative and also as active fibers. PCFs used as active fibers have been first reported in [6.1]. In particular, the possibility of obtaining very small– or very large–mode area with this new kind of optical fibers has been exploited to realize new fiber lasers [6.1, 6.2] or fiber amplifiers with single transverse mode operation and efficiency higher than in conventional doped fibers. Moreover, it has been investigated the influence of the PBG on the spontaneous emission of an $Er^{3+}$-doped PCF [6.3]. However, the research on rare earth-doped PCFs has been mainly focused on the development of $Yb^{3+}$-doped fiber lasers [6.4] and on the possibility offered by PCFs to properly control the overlap factors between the field and the dopant [6.5, 6.6], or to reduce the pump power in erbium-doped fiber amplifiers (EDFAs) [6.7]. The advantages offered by PCFs have been exploited to realize also cladding-pumped lasers and amplifiers. In particular, a highly efficient cladding-pumped single transverse mode $Yb^{3+}$-doped PCF laser has been demonstrated [6.8], while a large-mode-area $Nd^{3+}$-doped one has been proposed [6.9], as well as a wide-band cladding-pumped EDFA based on an air-clad PCF [6.10]. The first EDFA realized with a triangular PCF, providing up to 44 dB of internal gain, has been experimentally demonstrated in [6.11]. Then, the influence of the fiber length and of the wavelength on the amplifier gain and noise performances has been experimentally characterized [6.12]. Recently, the same small-core erbium-doped aluminosilicate PCF has been exploited to realize a simple Fabry Pérot laser with a slope efficiency of 57.3% and a threshold as low as 0.55 mW, and to demonstrate a device with a broadband tuning range, that is 104 nm around 1550 nm, and a laser threshold as low as 0.48 mW [6.13].

In order to study the amplification properties of the erbium-doped PCFs, a numerical model which combines the full-vector modal solver based on the FEM with a population and propagation rate equation solver, as reported in [6.5], has been developed. The FEM-based solver has been applied to evaluate the pump, the signal and the Amplified Spontaneous Emission (ASE) beam intensities, which are the input data for the population and propagation rate equations, describing the beam evolution along the doped fiber. These equations have been solved by means of the Runge-Kutta algorithm.

This amplifier model has been successfully applied to study the amplification properties of honeycomb PCFs and of a cobweb holey fiber [6.5, 6.14]. Results have demonstrated that active fibers with superior characteristics with respect to standard ones can be obtained by a proper PCF design. However, the PCFs considered in [6.5], due to the presence of the central air-hole in the honeycomb fibers and the very small core size of the cobweb one, sustain field distributions quite different from the fundamental mode of a standard optical fiber, and this can be critical in terms of coupling and splice losses.

The described EDFA model has been also applied to triangular erbium-doped PCFs providing numerical results in perfect agreement with experimental ones. Then, it has been used to design triangular PCFs which exhibit high gain values and low losses when spliced with a standard SMF [6.15]. With the triangular PCF EDFA here proposed it is no more necessary to use a high NA fiber to achieve good intermediate mode matching, as in [6.11]. Moreover, simulation results have demonstrated the practical application of erbium-doped PCFs as amplifiers and lasers compatible with conventional optical fiber systems.

## 6.1 Model for doped-fiber amplifiers

The overlap between the dopant and the field distributions provides a figure of the interaction between the dopant ions and the signals and, in turn, of the amount of the achievable amplification. This overlap can be easily evaluated in conventional doped-fiber amplifiers, since both the dopant concentration, often constant all over the fiber core, and the field profile are well known. PCFs, on the contrary, present a very complicated refractive index distribution, which makes difficult the field evaluation, unless proper numerical methods, able to accurately describe the local variation of the field, are adopted.

The field component profiles have been obtained by means of the full-vector modal solver. In particular, the normalized intensity mode distribution $i(x, y)$ has been derived according to Eq. (A.5), as described in Appendix A. By definition, the integral of the normalized intensity over the whole transverse PCF cross-section is equal to one. The FEM is applied to evaluate the pump, the signal and the ASE beam intensities, defined as

$$I_k(x, y, z) = i_k(x, y)P_k(z) , \tag{6.1}$$

with the subscript $k$ referring to the pump, the signal, or the ASE spectrum [6.16, 6.17]. These intensities are the input data for the population rate equations and the propagation rate equations, which describe the field—dopant interaction and the evolution of the pump, the signal, and the ASE beams along the doped fiber. These equations are solved by means of the Runge-Kutta algorithm.

It is worth noting that in the present analysis of the PCF-based EDFA performances a metastable lifetime equal to 10.5 ms has been assumed. Potentially, photonic crystals may alter the properties of active materials [6.18], increasing the lifetime of the rare earth elements incorporated into silica. At present time, due to the lack of experimental measurements, no data are available. However, the proposed approach can be indifferently applied whatever the lifetime. Finally, it is important to underline that the analysis has been here restricted to erbium as dopant, but the method can be applied to the study of any other rare earth ion.

## 6.2 EDFAs based on honeycomb and cobweb PCFs

Erbium-doped PCF performances have been analyzed, in order to understand how the air-hole geometry and the dopant distribution can be designed, with the aim to improve the amplification properties. Different PCF types have been considered, which guide light by exploiting the PBG effect or the modified TIR. All the simulations have been performed by applying the model previously described, which allows an accurate description of the amplification of WDM signals simultaneously propagating along the doped PCF.

First of all, the amplification properties of an erbium-doped PCF with the air-holes arranged in a honeycomb lattice have been considered. As reported in Fig. 6.1, an extra air-hole of radius $r_c$ has been introduced in the center of the fiber cross-section, acting as the defect which provides the light-guiding

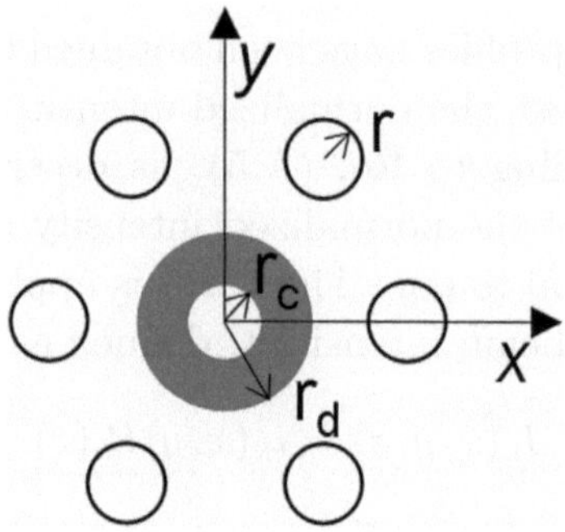


Figure **6.1**: Schematic of the central part of the honeycomb PCF cross-section, showing the doped region in grey [6.5].

through the PBG effect. The erbium dopant has been placed in a ring around the central air-hole with major radius $r_\mathrm{d}$ and minor radius $r_\mathrm{c}$, where the signal and the pump intensity distributions are more significant. In order to successfully use the proposed EDFA model with the honeycomb PCFs, it is necessary to calculate also the upper and the lower limit of the PBG, within which the core defect allows the field propagation. These have been evaluated using a freely available software package [6.19], and then the full-vector FEM-based solver has been applied by properly translating the guided-mode research inside the PBG. The gain performances of the honeycomb PCF EDFA have been analyzed by investigating how the radius $r_\mathrm{c}$ of the defect air-hole and the dopant concentration distribution on the fiber cross-section can be optimized to improve the amplification with respect to the standard step-index doped fibers.

The possibility offered by PCFs to properly control the guided-mode field distribution and, as a consequence, the overlap factor between the field and the dopant has been here exploited. Figure 6.2a and b report the normalized intensities along the $y$-axis of the pump at 980 nm and of the signal at 1560 nm for the honeycomb PCF with $r_\mathrm{c} = r$ and for a standard SMF. It is important to underline that the pump and the signal normalized intensity peaks are almost the same for the honeycomb PCF. On the contrary, the maximum signal intensity is only 50% of the pump one for the SMF. These differences in the field distributions significantly affect the overlap integrals, according to the value of the doped-area radius. Results have demonstrated that, by properly reducing the guided-mode area of the pump and the signal, with an

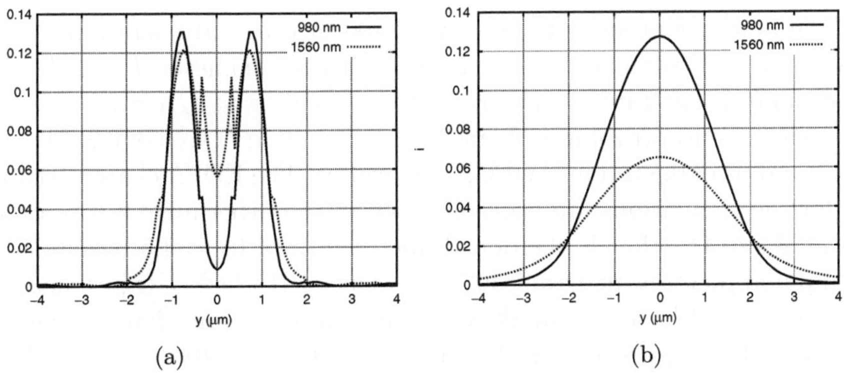


Figure **6.2**: Normalized intensities at 980 and 1560 nm of (a) the honeycomb PCF with $r_c = r$ and of (b) the standard SMF [6.5].

erbium-doped honeycomb PCF it is possible to achieve a gain enhancement respect to a SMF of more than 10 dB for a fixed dopant concentration per unit length [6.5, 6.14].

The amplification properties of a cobweb holey fiber, shown in Fig. 4.1, have been also analyzed, by changing the radius $r_d$ of the erbium-doped area in the center of the fiber cross-section. Results have confirmed the usefulness of the method here proposed in order to design doped-PCFs with better performances than conventional erbium-doped fibers.

## 6.3 EDFAs based on triangular PCFs

Both the honeycomb PCF and the cobweb holey fiber present high coupling losses toward standard SMFs, commonly used in telecommunication systems, as already stated. Consequently, it is not possible to completely exploit the advantages in term of signal gain provided by these erbium-doped PCF types.

As an interesting alternative, erbium-doped triangular PCFs can be considered. In particular, the amplification properties of the doped fiber used for the first experimental demonstration of a triangular PCF EDFA [6.11] have been extensively analyzed with the model here proposed. After making a comparison with the measurement results, it has been investigated the influence of the dopant radius and of the diameter of the air-holes in the first ring on the amplifier performances. Great attention has been payed to triangular PCF designs which allow to greatly reduce the splice losses toward the conventional SMFs.

The first erbium-doped fiber considered in the study here reported is the triangular PCF presented in [6.11]. The air-hole diameter $d$ and the pitch $\Lambda$ are equal to 1 and 2 μm, respectively, corresponding to $d/\Lambda = 0.5$. The erbium ions, whose concentration is about $2.6 \times 10^{25}$ ions/m$^3$, are confined in a region with a radius of 0.5 μm in the PCF core center. The EDFA characteristics chosen for the simulations are the same used in the experimental setup reported in Fig. 6.3 [6.11], that is the doped fiber length is $L_\mathrm{F} = 4.5$ m and the backward pump power at 980 nm is 225 mW. Due to the lack of experimental data, the background losses of the doped PCF have been considered equal to zero. However, this approximation does not affect the validity of the results, since the erbium-doped fiber length is usually only few meters, differently from the PCF-based Raman amplifiers, where the fiber losses represent a crucial factor in the amplifier design [6.20].

The amplifier performances have been calculated for different signal wavelengths, that is 1533, 1550, 1570, and 1590 nm, by changing the signal input power. In order to make a comparison with the experimental measurement results shown in Fig. 6.4a, the internal gain values calculated with the simulations are reported as a function of the power at the amplifier output in Fig. 6.4b. It is important to underline that the results here presented are in very good agreement with the experimental values obtained in [6.11], thus proving the validity of the model. The amplifier spectral gain and noise figure, evaluated for a single signal in the wavelength range between 1520 and 1580 nm, are reported in Fig. 6.5a. Notice that a signal internal gain of 46 dB has been reached at 1533 nm. Moreover, it is important to underline that

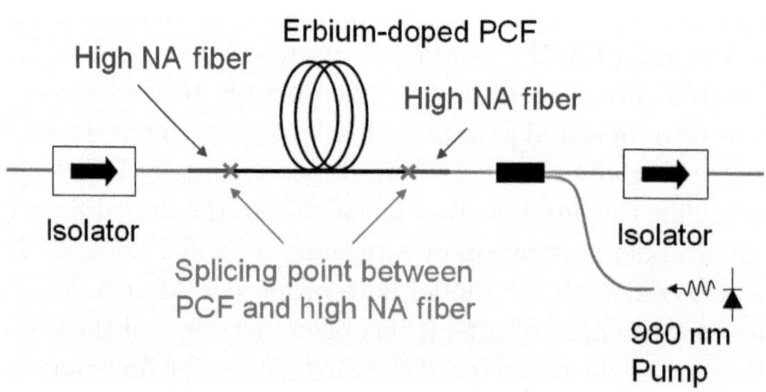


Figure **6.3**: Schematic diagram of the PCF-based EDFA proposed in [6.11].

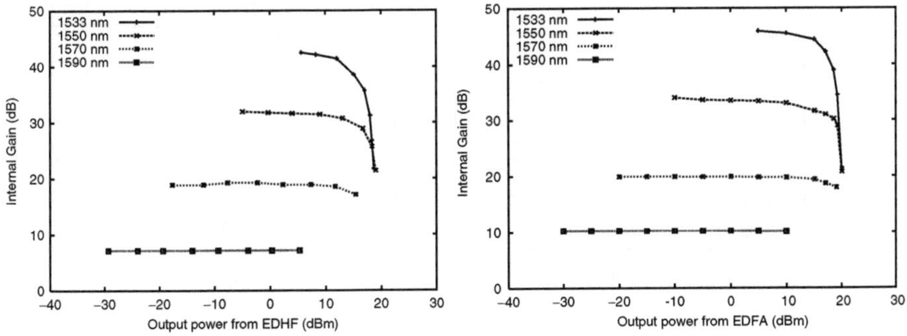


(a) (b)

Figure **6.4**: Single channel gain saturation characteristics of the EDFA at various wavelengths obtained with (a) the experimental measurements reported in [6.11] and (b) the numerical simulations.

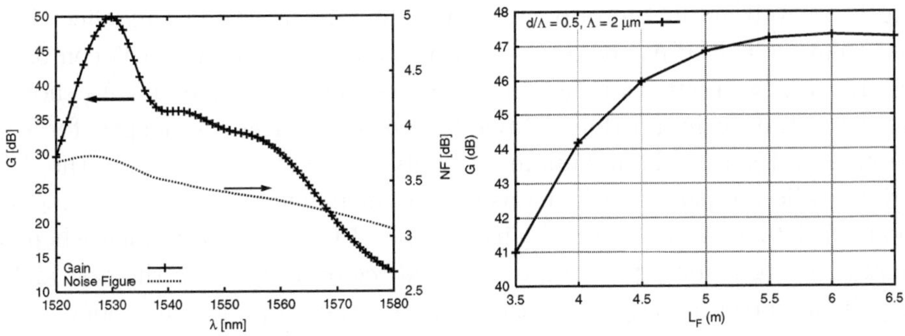


(a) (b)

Figure **6.5**: (a) Spectral gain and noise figure of the PCF-based amplifier. (b) Gain at 1533 nm versus $L_{\mathrm{F}}$ for the PCF with $d/\Lambda = 0.5$ and $\Lambda = 2$ μm [6.15].

the peak value of 49.5 dB has been obtained at 1530 nm, while the gain decreases to 33.5 dB at 1550 nm. In order to calculate the optimum length of the PCF-based EDFA experimentally realized, the amplifier gain for the signal at 1533 nm has been evaluated for different $L_{\mathrm{F}}$ values. As shown in Fig. 6.5b, a further gain increase of 1.5 dB can be achieved with a doped-fiber length of 5.5 m. In fact, a longer PCF allows to fully exploit the pump power, with a consequent gain increase.

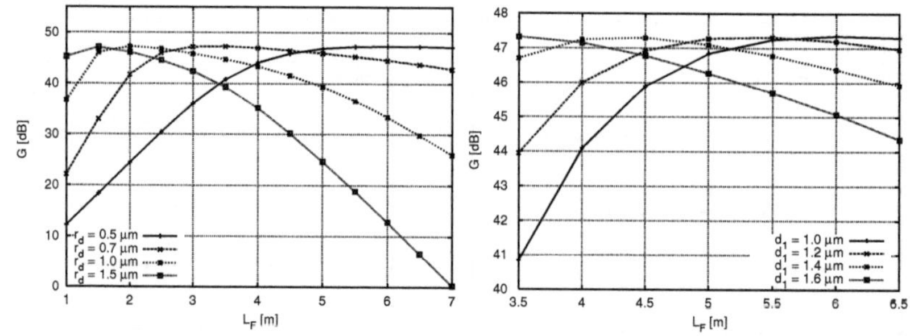


(a) (b)

Figure **6.6**: Gain versus the erbium-doped PCF length $L_{\mathrm{F}}$ for different (a) $r_{\mathrm{d}}$ values when $d_1 = d$ and (b) $d_1$ values when $r_{\mathrm{d}} = 0.5$ µm [6.15].

As in conventional doped fibers, the amplifier gain depends on the fiber length, as well as on the dopant radius. In order to study this effect in the triangular PCF EDFA, the dopant radius $r_{\mathrm{d}}$ has been varied between 0.5 µm and 1.5 µm, without changing the PCF geometric parameters. As shown in Fig. 6.6a, the optimum doped fiber length, defined as the length for which the gain at 1533 nm is maximum, strongly decreases by enlarging the doped region, while the peak gain remains almost unchanged. In particular, a gain of about 47 dB has been obtained with a doped PCF only 1.5 m long when $r_{\mathrm{d}}$ is equal to 1.5 µm, corresponding to a doped region tangent to the first air-hole ring. It has been already demonstrated that the amplifier gain depends also on the field confinement inside the erbium-doped PCF [6.5]. This confinement can be modified by changing only the diameter $d_1$ of the air-holes belonging to the first ring. In Fig. 6.6b the gain versus $L_{\mathrm{F}}$ is reported for $d_1$ between 1.0 and 1.6 µm, by keeping $r_{\mathrm{d}}$ fixed to 0.5 µm, in order to study only how the field confinement variation affects the gain. Results have shown that, for a fixed length of the doped PCF, the gain strongly depends on $d_1$. For example, by considering $L_{\mathrm{F}} = 3.5$ m, the gain is less than 41 dB when $d_1 = 1.0$ µm, while it is more than 47 dB when $d_1 = 1.6$ µm. This suggests that an enlarged first ring air-hole size can be usefully exploited in order to enhance the guided-mode field confinement, both at the signal and the pump wavelengths, and, consequently, to increase the overlap integrals [6.5]. However, the maximum gain obtainable seems to be almost independent on $d_1$, provided that the fiber length is properly adjusted. When both $r_{\mathrm{d}}$ and $d_1$ are different from the initial values, in particular $r_{\mathrm{d}} = 1.2$ µm and $d_1 = 1.6$ µm, the same maximum gain is obtained with $L_F = 1.5$ m. Simulations have been performed also to evaluate the role of the second air-hole ring size $d_2$, but results have demonstrated that the value of $d_2$ does not affect in any way the amplifier performances.

Results presented so far suggest that high gain values can be easily obtained with erbium-doped PCFs. However, the losses due to the mode mismatch between the PCF and the standard SMF can be a severe limiting factor. In [6.11] this problem has been partially solved by using a high NA fiber as intermediate fiber, as shown in the experimental setup of Fig. 6.3. Each splice loss between the PCF and the high NA fiber has been measured to be about 1.7 dB. Moreover, also the loss between the high NA fiber and the standard one should be included, and this corresponds to an overall loss of almost 3 dB. Instead of using an intermediate fiber, a possible solution could be the design of a PCF which minimizes the coupling loss. In literature it has been already shown that the splice losses decrease by increasing the pitch $\Lambda$ and by reducing $d/\Lambda$ [6.21]. Thus, a triangular PCF with $d/\Lambda = 0.5$, $r_d/\Lambda = 0.25$ and $\Lambda = 3$ μm, instead of $\Lambda = 2$ μm as in [6.11], has been considered. According to [6.22], this fiber is single mode both at the pump and the signal wavelengths. The splice losses due to the mode field mismatch have been calculated on the basis of formulae proposed in [6.23] and [6.24]. This calculation is not enough for a very accurate description of the PCF splicing issue, which is not as simple as in conventional fibers [6.25], but it is effective to the present analysis aim. The calculated loss between a SMF-28 fiber and the triangular PCF is 5.8 and 3.6 dB for $\Lambda = 2$ and 3 μm, respectively. Notice that the coupling losses are high for the PCF with $d/\Lambda = 0.5$ and $\Lambda = 2$ μm, being its effective area at 1550 nm, that is 6.57 $\mu m^2$, more than one order of magnitude lower than that of the SMF-28, that is 86 $\mu m^2$. It is important to underline that the losses are reduced of about 2 dB when the pitch increases to 3 μm, while the maximum gain is still 47.5 dB for both the PCFs.

This result would suggest to greatly increase the pitch, but the triangular PCFs could become multi-mode, according to [6.22]. To avoid this drawback and to further decrease the losses, the ratio $d/\Lambda$ has been reduced to 0.4 and $\Lambda$ values in the range 2—12 μm have been considered. It is important to underline that $r_d$ has been kept fixed to $0.25\Lambda$, so the doped area enlarges proportionally with the whole PCF cross-section as $\Lambda$ increases. The signal gain at 1533 nm is reported in Fig. 6.7a as a function of the amplifier length for the different PCFs with $d/\Lambda = 0.4$. Notice that the optimum length of the PCF-based EDFA is always between 5.5 and 6 m, regardless of the pitch value, even if the gain slightly decreases for larger $\Lambda$. However, the maximum gain is still 45.5 dB for the PCF with $\Lambda = 12$ µm. Notice that the PCFs with $d/\Lambda = 0.4$ and $\Lambda$ in the range 3–12 µm exhibit the same gain dependence on the fiber length. Only the PCFs with $\Lambda = 2$ and 2.5 µm show a slightly different behavior. This is due to the difference between the overlap integrals at the signal and the pump wavelengths, which becomes lower when the pitch $\Lambda$ increases, as reported in Fig. 6.7b. Results reported in Fig. 6.7a demonstrate that no detriment to the amplifier gain is caused by the pitch enlargement in the erbium-doped triangular PCF. On the contrary, this provides a significant splice loss reduction, as reported in Fig. 6.8a, which shows the calculated coupling losses between a SMF-28 and a triangular PCF versus the pitch $\Lambda$. Notice that for the PCFs with $d/\Lambda = 0.4$ the splice losses decrease from 4.5 dB, when $\Lambda = 2$ µm, to only 0.003 dB, when $\Lambda = 8$ µm. This great reduction can be explained by considering that the PCF with $\Lambda = 8$ µm, which yields a maximum gain of 46.5 dB for $L_F = 4.5$ m, has an effective area of 91 $\mu m^2$, very similar to the SMF-28 one. Since the guided-mode field confinement decreases as the pitch value becomes higher, the effective area of the PCFs with $\Lambda > 8$ µm is larger than that of the SMF-28, thus causing a worsening of the coupling loss. However, the difference between the effective area of the triangular PCF and of the SMF-28 increases slowly with $\Lambda$. For example, even if its effective area is more than twice the one of the SMF-28, that is 198 $\mu m^2$, the splice losses of the PCF with $\Lambda = 12$ µm are lower than those of

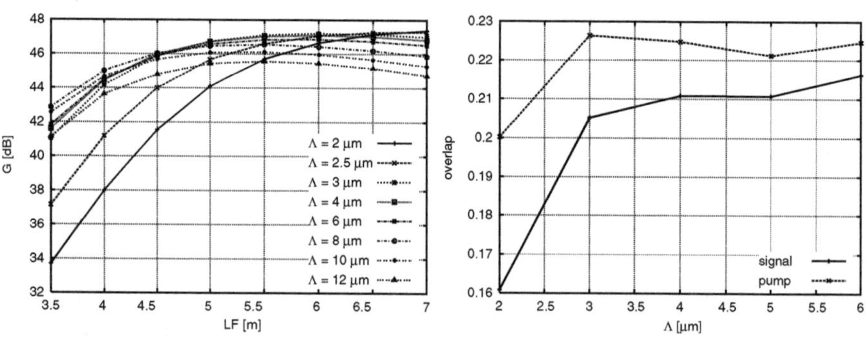


(a) (b)

Figure **6.7**: (a) Gain of the PCF with $d/\Lambda = 0.4$ for different $\Lambda$ values. (b) Overlap integrals at the signal and the pump wavelengths for $\Lambda$ between 2 and 6 µm [6.15].

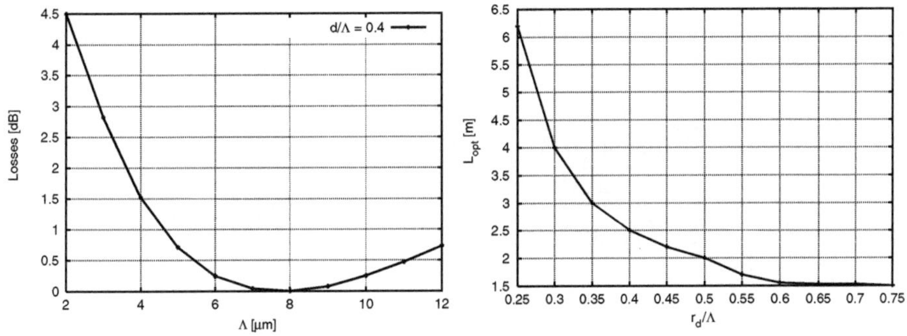


(a) (b)

Figure **6.8**: (a) Loss between a SMF-28 fiber and a triangular PCF with $d/\Lambda = 0.4$ for different $\Lambda$ values. (b) Optimum fiber length of the PCF with $d/\Lambda = 0.4$ and $\Lambda = 6$ μm for different $r_d/\Lambda$ values [6.15].

the erbium-doped fiber presented in [6.11]. In conclusion, results here reported have shown that with the triangular PCF with $d/\Lambda = 0.4$ and $\Lambda = 8$ μm it is possible to reduce the coupling loss of almost 5 dB, while keeping unchanged the amplifier gain performances.

Finally, the PCF with $d/\Lambda = 0.4$ and $\Lambda = 6$ μm, which provides 0.25 dB loss toward the standard SMF, as shown in Fig. 6.8a, has been considered in detail and simulations have been performed to evaluate the effect of the dopant radius on the amplifier gain characteristics. In particular, the radius $r_d$ has been changed between $0.25\Lambda$ and $0.75\Lambda$. Here, $d_1$ has been kept unchanged, because enlarging $d_1$ would cause an increase of the guided-mode field confinement, with a consequent growth of the splice losses. Results are reported in Fig. 6.8b, where the optimum doped fiber length $L_{opt}$ versus $r_d/\Lambda$ is shown. Notice that $L_{opt}$ decreases from 6.2 m, when $r_d = 0.25\Lambda$, to 1.5 m, when $r_d = 0.75\Lambda$, while the maximum gain remains higher than 47 dB. In conclusion, an example of the design parameters for two of the proposed PCF amplifiers with a desired gain of 47 dB is reported in Table 6.1.

In the last part of the present analysis a coupling loss reduction between the PCF-based EDFA and the conventional SMF has been obtained by designing an erbium-doped PCF with a larger effective area. To this aim, it has been considered again the triangular fiber realized in [6.11] and shown on the left in the inset of Fig. 6.9. The first air-hole ring has been removed from the fiber cross-section, as reported on the right in the inset of Fig. 6.9, thus obtaining

Table 6.1: Design parameters for two of the proposed PCF amplifiers [6.15].

| | $\mathbf{\Lambda = 2}$ **μm** | $\mathbf{\Lambda = 6}$ **μm** |
|---|---|---|
| $d/\Lambda$ | 0.5 | 0.4 |
| **Losses (dB)** | 5.8 | 0.25 |
| $r_d$ (μm) | 1.5 | 4.5 |
| $L_{opt}$ (m) | 1.5 | 1.5 |

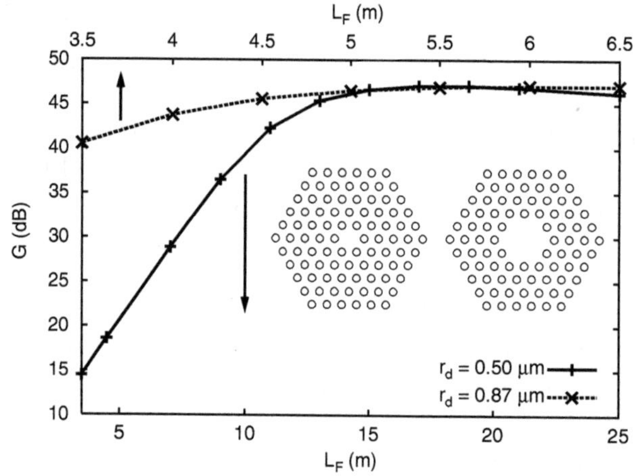


Figure **6.9**: Gain at 1533 nm as a function of $L_F$ for the PCF without the first air-hole ring for two different $r_d$ values. *Inset*: triangular PCF cross-section (*left*) with and (*right*) without the first air-hole ring.

a larger silica core region and a lower field confinement. Without the first air-hole ring, the effective area of the PCF with $d/\Lambda = 0.5$ and $\Lambda = 2$ μm becomes almost 19 μm$^2$, that is about 3.6 times higher than that of the doped-fiber presented in [6.11]. The effective area increase, clearly shown in Fig. 6.10, where the fundamental component of the guided-mode magnetic field at 1533 nm of the triangular PCF with and without the first air-hole ring is reported, causes a coupling loss reduction of 3.5 dB with respect to the 5.8 dB of the erbium-doped PCF presented in [6.11]. By taking into account the gain behavior as a function of the length $L_F$ of the doped PCF without the first air-hole ring, reported in Fig. 6.9, it is possible to notice that a maximum gain of about 47 dB can still be reached, providing that a longer EDFA, that is 19 m long, is considered. This conclusion is a direct result of the lower overlap

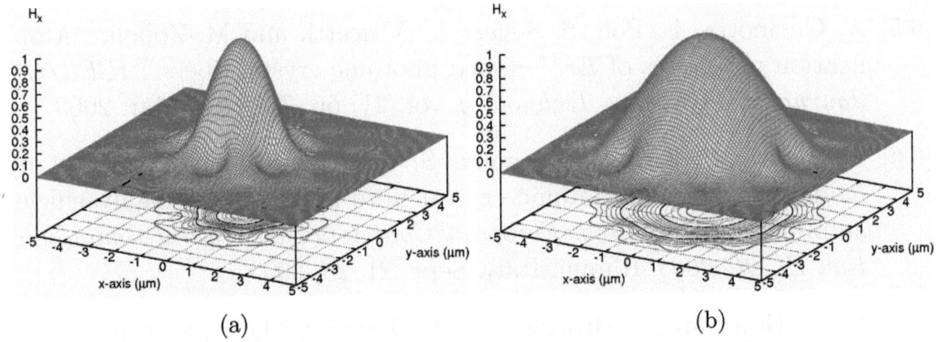


Figure **6.10**: Fundamental component of the guided-mode magnetic field at 1533 nm of the triangular PCF (a) with and (b) without the first air-hole ring.

integral value at both the pump and the signal wavelengths. Then the radius of the erbium-doped area has been changed in the triangular PCF without the first air-hole ring, in order to obtain the same maximum gain with a shorter fiber. Simulation results reported in Fig. 6.9 have shown that, by choosing $r_\mathrm{d} = 0.435\Lambda = 0.87$ μm, a maximum gain of about 47 dB has been obtained with $L_\mathrm{F} = 6$ m.

# Bibliography

[6.1] W. J. Wadsworth, J. C. Knight, W. H. Reeves, P. St. J. Russell, and J. Arriaga, "$Yb^{3+}$-doped photonic crystal fibre laser," *Electronics Letters*, vol. 36, pp. 1452–1454, Aug. 2000.

[6.2] T. Søndergaard, "Photonic crystal distributed feedback fiber lasers with Bragg gratings," *IEEE/OSA Journal of Lightwave Technology*, vol. 18, pp. 589–597, Apr. 2000.

[6.3] R. F. Cregan, J. C. Knight, P. St. J. Russell, and P. J. Roberts, "Distribution of spontaneous emission from and $Er^{3+}$-doped photonic crystal fiber," *IEEE/OSA Journal of Lightwave Technology*, vol. 17, pp. 2138–2141, Nov. 1999.

[6.4] K. Furusawa, T. M. Monro, P. Petropoulos, and D. J. Richardson, "Modelocked laser based on ytterbium doped holey fibre," *Electronics Letters*, vol. 37, pp. 560–561, Apr. 2001.

[6.5] A. Cucinotta, F. Poli, S. Selleri, L. Vincetti, and M. Zoboli, "Amplification properties of $Er^{3+}$-doped photonic crystal fibers," *IEEE/OSA Journal of Lightwave Technology*, vol. 21, pp. 782–788, Mar. 2003.

[6.6] S. Hilaire, P. Roy, D. Pagnoux, S. Février, and D. Bayart, "Large mode $Er^{3+}$-doped photonic crystal fibre amplifier for highly efficient amplification," in *Proc. European Conference on Optical Communication ECOC 2003*, Rimini, Italy, Sept. 21–25, 2003.

[6.7] K. G. Hougaard, J. Broeng, and A. Bjarklev, "Low pump power photonic crystal fibre amplifiers," *Electronics Letters*, vol. 39, pp. 599–600, Apr. 2003.

[6.8] K. Furusawa, A. Malinowski, J. H. Price, T. M. Monro, J. K. Sahu, J. Nilsson, and D. J. Richardson, "Cladding pumped Ytterbium-doped fiber laser with holey inner and outer cladding," *Optics Express*, vol. 9, pp. 714–720, Dec. 2001. Available at: http://www.opticsexpress.org/abstract.cfm?URI=OPEX-9-13-714

[6.9] P. Glas and D. Fischer, "Cladding pumped large-mode-area $Nd$-doped holey fiber laser," *Optics Express*, vol. 10, pp. 286–290, Mar. 2002. Available at: http://www.opticsexpress.org/abstract.cfm?URI=OPEX-10-6-286

[6.10] C. Simonneau, P. Bousselet, G. Melin, L. Provost, and C. Moreau, "High-power air-clad photonic crystal fiber cladding-pumped EDFA for WDM applications in the C-band," in *Proc. European Conference on Optical Communication ECOC 2003*, Rimini, Italy, Sept. 21–25, 2003.

[6.11] T. Kogure, K. Furusawa, J. H. Lee, T. M. Monro, and D. J. Richardson, "An erbium doped holey fiber amplifier and ring laser," in *Proc. European Conference on Optical Communication ECOC 2003*, Rimini, Italy, Sept. 21–25, 2003, paper post-deadline.

[6.12] K. Furusawa, T. Kogure, T. M. Monro, and D. J. Richardson, "High gain efficiency amplifier based on an erbium doped aluminosilicate holey fiber," *Optics Express*, vol. 12, pp. 3452–3458, July 2004. Available at: http://www.opticsinfobase.org/abstract.cfm?URI=oe-12-15-3452

[6.13] K. Furusawa, T. Kogure, J. K. Sahu, J. H. Lee, T. M. Monro, and D. J. Richardson, "Efficient low-threshold lasers based on an erbium-doped holey fiber," *IEEE Photonics Technology Letters*, vol. 17, pp. 25–27, Jan. 2005.

[6.14] S. Selleri, A. Cucinotta, F. Poli, L. Vincetti, and M. Zoboli, "Amplification properties of erbium doped photonic crystal fibers," in *Proc. European Conference on Optical Communication ECOC 2002*, Copenhagen, Denmark, Sept. 8–12, 2002.

[6.15] A. Cucinotta, F. Poli, and S. Selleri, "Design of erbium-doped triangular photonic crystal fiber based amplifiers," *IEEE Photonics Technology Letters*, vol. 16, pp. 2027–2029, Sept. 2004.

[6.16] F. D. Pasquale and M. Zoboli, "Analysis of erbium doped waveguide amplifiers by a full-vectorial finite-element method," *IEEE/OSA Journal of Lightwave Technology*, vol. 11, pp. 1565–1574, Oct. 1993.

[6.17] C. R. Giles and E. Desurvire, "Modeling erbium-doped fiber amplifiers," *IEEE/OSA Journal of Lightwave Technology*, vol. 9, pp. 271–283, Feb. 1991.

[6.18] T. Søndergaard and B. Tromborg, "General theory for spontaneous emission in active dielectric microstructures: example of a fiber amplifier," *Physical Review A*, vol. 64, pp. 033 812–1–033 812–14, Sept. 2001.

[6.19] S. G. Johnson and J. D. Joannopoulos, "Block-iterative frequency-domain methods for Maxwell's equations in a planewave basis," *Optics Express*, vol. 8, pp. 173–179, Jan. 2001. Available at: http://www.opticsexpress.org/abstract.cfm?URI=OPEX-8-3-173

[6.20] C. J. S. de Matos, K. P. Hansen, and J. R. Taylor, "Experimental characterisation of Raman gain efficiency of holey fibre," *Electronics Letters*, vol. 39, pp. 424–425, Mar. 2003.

[6.21] J. T. Lizier and G. E. Town, "Splice losses in holey fibers," *IEEE Photonics Technology Letters*, vol. 13, pp. 794–796, Aug. 2001.

[6.22] N. A. Mortensen, J. R. Folkenberg, M. D. Nielsen, and K. P. Hansen, "Modal cutoff and the V parameter in photonic crystal fibers," *Optics Letters*, vol. 28, pp. 1879–1881, Oct. 2003.

[6.23] N. A. Mortensen, "Effective area of photonic crystal fiber," *Optics Express*, vol. 10, pp. 341–348, Apr. 2002. Available at: http://www.opticsexpress.org/abstract.cfm?URI=OPEX-10-7-341

[6.24] J. H. Chong and M. K. Rao, "Development of a system for laser splicing photonic crystal fiber," *Optics Express*, vol. 11, pp. 1365–1370, May 2003. Available at: http://www.opticsexpress.org/abstract.cfm?URI=OPEX-11-12-1365

[6.25] B. Bourliaguet, C. Paré, F. Émond, A. Croteau, A. Proulx, and R. Vallée, "Microstructured fiber splicing," *Optics Express*, vol. 11, pp. 3412–3417, Dec. 2003. Available at: http://www.opticsexpress.org/abstract.cfm?URI=OPEX-11-25-3412

# Appendix A

# Finite Element Method

## A.1 Formulation

All the analyses of the PCF properties presented in this book have been performed by using the FEM. The FEM allows the PCF cross-section in the transverse $x - y$ plane to be divided into a patchwork of triangular elements, which can be of different sizes, shapes, and refractive indices. In this way any kind of geometry, including the PCF air-holes, as well as the medium characteristics, can be accurately described. In particular, the FEM is suited for studying fibers with nonperiodic air-hole arrangements. Moreover, it provides a full-vector analysis which is necessary to model PCFs with large air-holes and high index variations, and to accurately predict their properties [A.1].

The formulation of the FEM here considered is based on the curl–curl equation. For a medium described by the complex tensors of the relative dielectric permittivity $\bar{\bar{\epsilon}}_r$ and the magnetic permeability $\bar{\bar{\mu}}_r$ it reads

$$\overline{\nabla} \times (\bar{\bar{\varepsilon}}_r^{-1} \overline{\nabla} \times \overline{h}) - k_0^2 \bar{\bar{\mu}}_r \overline{h} = 0 \;, \qquad \text{(A.1)}$$

where $\overline{h}$ is the magnetic field, and $k_0 = 2\pi/\lambda$ is the wave number in the vacuum, $\lambda$ being the wavelength. The magnetic field of the modal solution is expressed as $\overline{h} = \overline{H} e^{-\gamma z}$, where $\overline{H}$ is the field distribution on the transverse plane and

$$\gamma = \alpha + j k_0 n_{\text{eff}} \qquad \text{(A.2)}$$

is the complex propagation constant, with $\alpha$ the attenuation constant and $n_{\text{eff}}$ the effective index.

By applying the variational finite element procedure, Eq. (A.1) yields the algebraic problem [A.2]

$$([A] - (\frac{\gamma}{k_0})^2[B])\{H\} = 0 \ , \tag{A.3}$$

where the eigenvector $\{H\}$ is the discretized magnetic field-vector distribution of the mode. The matrices $[A]$ and $[B]$ are sparse and symmetric, thus allowing an efficient resolution of Eq. (A.3) by means of high-performance algebraic solvers. In order to enclose the computational domain without affecting the numerical solution, anisotropic Perfectly Matched Layers (PML) are placed before the outer boundary [A.3, A.4]. This formulation is able to deal with anisotropic material both in terms of dielectric permittivity and magnetic permeability, allowing anisotropic PML to be directly implemented.

The FEM has allowed the successful investigation of PCF dispersion [A.5–A.8], amplification [A.9, A.10] and nonlinear properties [A.11–A.13]. Moreover, the complex FEM formulation has been very useful, for instance, to evaluate the PCF leakage or confinement losses, due to the finite number of air-hole rings in the cladding lattice [A.3, A.14].

In addition, the high flexibility of the method results in solutions whose accuracy has been thoroughly checked, either considering different FEM formulations or through comparisons with different numerical approaches [A.15, A.16]. Furthermore, fiber symmetry can be used to reduce the computational domain and, consequently, both time and memory required, without affecting the accuracy of the computed solution.

As an example, a PCF cross-section and the corresponding mesh used for the simulations are reported in Fig. A.1a and b, respectively. Notice that, by properly changing the dimension of the triangular elements which constitute the mesh, it is possible to accurately describe all the regions with different geometric and dielectric properties in the fiber transverse section. In particular, as shown in Fig. A.1b, the silica region, where the guided-mode field is mainly confined, are described with a lot of triangles of reduced dimensions. The magnetic field fundamental component of the guided mode, computed at 1550 nm, is reported in Fig. A.1c.

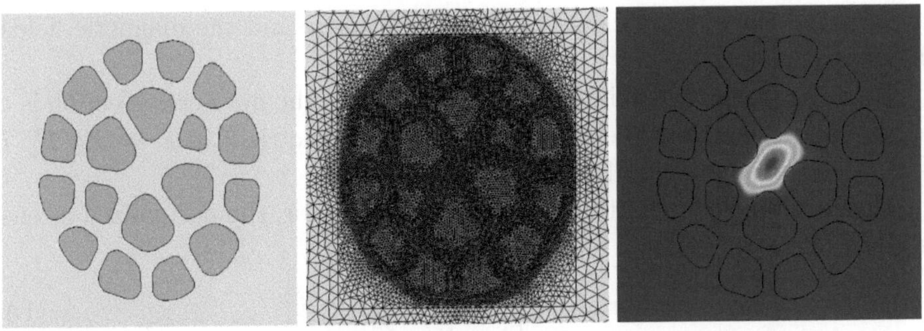

(a) (b) (c)

Figure **A.1**: (a) Geometry and (b) mesh of the cross-section of a small-core nonlinear PCF. Green and yellow regions represent, respectively, the air-holes and the silica bulk. (c) Fundamental component of the magnetic field at 1550 nm evaluated with the FEM-based full-vector modal solver.

## A.2 PCF parameter evaluation

### Dispersion

Starting from the knowledge of the effective refractive index $n_{\text{eff}}$ versus the wavelength, the dispersion parameter

$$D(\lambda) = -\frac{\lambda}{c}\frac{d^2 n_{\text{eff}}}{d\lambda^2} \tag{A.4}$$

can be derived using simple finite difference formulae. The chromatic dispersion of silica is taken into account through the Sellmeier equation [A.17], so the refractive index of the structure is changed, according to the working wavelength, before using the FEM solver to get the modal field and $n_{\text{eff}}$, as in Eq. (A.2).

### Nonlinear coefficient

The FEM can be exploited to evaluate the guided-mode field distribution in PCFs, necessary to compute the effective area and the nonlinear coefficient.

In order to accurately evaluate the effective area, the fundamental mode intensity distribution is calculated from the Poynting vector definition, which involves the three components of both the electric and the magnetic fields of the guided mode.

First, the magnetic field $\overline{H} = (H_x, H_y, H_z)$ on the fiber cross-section is calculated and then, from the expression of $\overline{H}$, the electric field $\overline{E} = (E_x, E_y, E_z)$ is obtained through the Maxwell equation.

Hence, from the definition of the Poynting vector, the normalized intensity is given by

$$i(x, y) = \frac{1}{P}\mathcal{R}e\left[\frac{\overline{E} \times \overline{H}^*}{2} \cdot \hat{z}\right] , \tag{A.5}$$

where $P$ is the integral of the intensity over the section of the PCF, that is,

$$P = \iint_S \mathcal{R}e\left[\frac{\overline{E} \times \overline{H}^*}{2} \cdot \hat{z}\right] dx\, dy = \iint_S \mathcal{R}e\left[\frac{E_x\, H_y^* - E_y^* H_x}{2}\right] dx\, dy\ . \tag{A.6}$$

Then, the effective area of the PCF fundamental guided mode can be calculated according to

$$A_{\text{eff}} = \frac{1}{\iint_S i^2(x, y) dx dy} , \tag{A.7}$$

where $i(x, y)$ is the guided-mode normalized intensity distribution, as in Eq. (A.5) [A.9].

As a consequence, the nonlinear coefficient can be evaluated as

$$\gamma = (2\pi/\lambda) \cdot \iint_S n_2(x, y) i^2(x, y) dx dy , \tag{A.8}$$

where $n_2(x, y)$ is $3 \cdot 10^{-20}$ $m^2/W$ in the silica bulk and 0 in the air-holes, and $i(x, y)$ is the normalized intensity, according Eq. (A.5) [A.9].

The accuracy of the $A_{\text{eff}}$ calculation here presented has been checked by comparing the values calculated with the FEM simulations with those experimentally measured with a Scanning Near-field Optical Microscope (SNOM). The SNOM technique can be used to evaluate the effective area of an optical fiber, since it permits to study the field distribution on its transverse section. In fact, an optical probe, that is a nanometric tapered single-mode optical fiber, is approached in the near field of the fiber under investigation. The image process is based on a pixel by pixel acquisition sequence, moving step by step the probe above the fiber cross-section and scanning all the region of interest. The image processing is performed by a computer, which stores all the data collected from each pixel [A.18]. It is important to underline that a good agreement has been found with the simulation results obtained through the FEM solver for all the PCFs considered in the measurements [A.19].

### Confinement losses

In a PCF with an infinite number of air-holes in the photonic crystal cladding, the propagation is theoretically lossless. However, in the fabricated fibers the number of air-holes is finite, so the guided modes are leaky.

The confinement loss CL of the mode is deduced from the attenuation constant $\alpha$ in Eq. (A.2) as

$$\mathrm{CL} = 20\alpha \log_{10} e = 8.686\alpha \quad (\mathrm{dB/m}) \ . \tag{A.9}$$

## Bibliography


[A.1] T. M. Monro, D. J. Richardson, N. G. R. Broderick, and P. J. Bennett, "Modeling large air fraction holey optical fibers," *IEEE/OSA Journal of Lightwave Technology*, vol. 18, pp. 50–56, Jan. 2000.

[A.2] S. Selleri, L. Vincetti, A. Cucinotta, and M. Zoboli, "Complex FEM modal solver of optical waveguides with PML boundary conditions," *Optical and Quantum Electronics*, vol. 33, pp. 359–371, Apr. 2001.

[A.3] D. Ferrarini, L. Vincetti, M. Zoboli, A. Cucinotta, and S. Selleri, "Leakage properties of photonic crystal fibers," *Optics Express*, vol. 10, pp. 1314–1319, Nov. 2002. Available at: http://www.opticsexpress.org/abstract.cfm?URI=OPEX-10-23-1314

[A.4] A. Cucinotta, G. Pelosi, S. Selleri, L. Vincetti, and M. Zoboli, "Perfectly matched anisotropic layers for optical waveguides analysis through the finite element beam propagation method," *Microwave and Optical Technology Letters*, vol. 23, pp. 67–69, Oct. 1999.

[A.5] F. Poli, A. Cucinotta, M. Fuochi, S. Selleri, and L. Vincetti, "Characterization of microstructured optical fibers for wideband dispersion compensation," *Journal of Optical Society of America A*, vol. 20, pp. 1958–1962, Oct. 2003.

[A.6] A. Cucinotta, S. Selleri, L. Vincetti, and M. Zoboli, "Holey fiber analysis through the finite-element method," *IEEE Photonics Technology Letters*, vol. 14, pp. 1530–1532, Nov. 2002.

[A.7] F. Poli, A. Cucinotta, S. Selleri, and A. H. Bouk, "Tailoring of flattened dispersion in highly nonlinear photonic crystal fibers," *IEEE Photonics Technology Letters*, vol. 16, pp. 1065–1067, Apr. 2004.

[A.8] A. Cucinotta, S. Selleri, L. Vincetti, and M. Zoboli, "Perturbation analysis of dispersion properties in photonic crystal fibers through the Finite Element Method," *IEEE/OSA Journal of Lightwave Technology*, vol. 20, pp. 1433–1442, Aug. 2002.

[A.9] A. Cucinotta, F. Poli, S. Selleri, L. Vincetti, and M. Zoboli, "Amplification properties of $Er^{3+}$-doped photonic crystal fibers," *IEEE/OSA Journal of Lightwave Technology*, vol. 21, pp. 782–788, Mar. 2003.

[A.10] A. Cucinotta, F. Poli, and S. Selleri, "Design of erbium-doped triangular photonic crystal fiber based amplifiers," *IEEE Photonics Technology Letters*, vol. 16, pp. 2027–2029, Sept. 2004.

[A.11] F. Poli, F. Adami, M. Foroni, L. Rosa, A. Cucinotta, and S. Selleri, "Optical parametric amplification in all-silica triangular-core photonic crystal fibers," *Applied Physics B*, vol. 81, pp. 251–255, July 2005.

[A.12] M. Fuochi, F. Poli, S. Selleri, A. Cucinotta, and L. Vincetti, "Study of Raman amplification properties in triangular photonic crystal fibers," *IEEE/OSA Journal of Lightwave Technology*, vol. 21, pp. 2247–2254, July 2003.

[A.13] M. Bottacini, F. Poli, A. Cucinotta, and S. Selleri, "Modeling of photonic crystal fiber Raman amplifiers," *IEEE/OSA Journal of Lightwave Technology*, vol. 22, pp. 1707–1713, July 2004.

[A.14] L. Vincetti, "Confinement losses in honeycomb fibers," *IEEE Photonics Technology Letters*, vol. 16, pp. 2048–2050, Sept. 2004.

[A.15] S. Selleri and M. Zoboli, "Performance comparison of finite element approaches for electromagnetic waveguides," *Journal of Optical Society of America A*, vol. 14, pp. 1460–1466, July 1997.

[A.16] S. Selleri and J. Petracek, "Modal analysis of rib waveguide through finite element and mode matching methods," *Optical and Quantum Electronics*, vol. 33, pp. 373–386, Apr. 2001.

[A.17] G. P. Agrawal, *Nonlinear Fiber Optics.* New York: Academic, 2001.

[A.18] M. Foroni, M. Bottacini, F. Poli, S. Selleri, and A. Cucinotta, "Scanning near-field optical microscope for characterization of single mode fibers," in *Proc. Optical Fibre Sensors Conference OFS-17*, Bruges, Belgium, May 23–27, 2005.

[A.19] M. Foroni, M. Bottacini, F. Poli, S. Selleri, and A. Cucinotta, "Effective area measurement of photonic crystal fibers through Scanning near-field optical microscope," in *Proc. ICONIC 2005*, Barcelona, Spain, June 8–10, 2005.

# Index

Springer Series in

# MATERIALS SCIENCE

*Editors:* R. Hull R. M. Osgood, Jr. J. Parisi H. Warlimont

40 **Reference Materials in Analytical Chemistry**
A Guide for Selection and Use
Editor: A. Zschunke

41 **Organic Electronic Materials**
Conjugated Polymers and Low Molecular Weight Organic Solids
Editors: R. Farchioni and G. Grosso

42 **Raman Scattering in Materials Science**
Editors: W. H. Weber and R. Merlin

43 **The Atomistic Nature of Crystal Growth**
By B. Mutaftschiev

44 **Thermodynamic Basis of Crystal Growth**
*P–T–X* Phase Equilibrium and Non-Stoichiometry
By J. Greenberg

45 **Thermoelectrics**
Basic Principles and New Materials Developments
By G. S. Nolas, J. Sharp, and H. J. Goldsmid

46 **Fundamental Aspects of Silicon Oxidation**
Editor: Y. J. Chabal

47 **Disorder and Order in Strongly Nonstoichiometric Compounds**
Transition Metal Carbides, Nitrides and Oxides
By A. I. Gusev, A. A. Rempel, and A. J. Magerl

48 **The Glass Transition**
Relaxation Dynamics in Liquids and Disordered Materials
By E. Donth

49 **Alkali Halides**
A Handbook of Physical Properties
By D. B. Sirdeshmukh, L. Sirdeshmukh, and K. G. Subhadra

50 **High-Resolution Imaging and Spectrometry of Materials**
Editors: F. Ernst and M. Rühle

51 **Point Defects in Semiconductors and Insulators**
Determination of Atomic and Electronic Structure from Paramagnetic Hyperfine Interactions
By J.-M. Spaeth and H. Overhof

52 **Polymer Films with Embedded Metal Nanoparticles**
By A. Heilmann

53 **Nanocrystalline Ceramics**
Synthesis and Structure
By M. Winterer

54 **Electronic Structure and Magnetism of Complex Materials**
Editors: D.J. Singh and D. A. Papaconstantopoulos

55 **Quasicrystals**
An Introduction to Structure, Physical Properties and Applications
Editors: J.-B. Suck, M. Schreiber, and P. Häussler

56 **$SiO_2$ in Si Microdevices**
By M. Itsumi

57 **Radiation Effects in Advanced Semiconductor Materials and Devices**
By C. Claeys and E. Simoen

58 **Functional Thin Films and Functional Materials**
New Concepts and Technologies
Editor: D. Shi

59 **Dielectric Properties of Porous Media**
By S.O. Gladkov

60 **Organic Photovoltaics**
Concepts and Realization
Editors: C. Brabec, V. Dyakonov, J. Parisi and N. Sariciftci

61 **Fatigue in Ferroelectric Ceramics and Related Issues**
By D.C. Lupascu

62 **Epitaxy**
Physical Principles and Technical Implementation
By M.A. Herman, W. Richter, and H. Sitter

Springer Series in
# MATERIALS SCIENCE

*Editors:* R. Hull   R. M. Osgood, Jr.   J. Parisi   H. Warlimont

63 **Fundamentals of Ion-Irradiated Polymers**
By D. Fink

64 **Morphology Control of Materials and Nanoparticles**
Advanced Materials Processing and Characterization
Editors: Y. Waseda and A. Muramatsu

65 **Transport Processes in Ion-Irradiated Polymers**
By D. Fink

66 **Multiphased Ceramic Materials**
Processing and Potential
Editors: W.-H. Tuan and J.-K. Guo

67 **Nondestructive Materials Characterization**
With Applications to Aerospace Materials
Editors: N.G.H. Meyendorf, P.B. Nagy, and S.I. Rokhlin

68 **Diffraction Analysis of the Microstructure of Materials**
Editors: E.J. Mittemeijer and P. Scardi

69 **Chemical-Mechanical Planarization of Semiconductor Materials**
Editor: M.R. Oliver

70 **Applications of the Isotopic Effect in Solids**
By V.G. Plekhanov

71 **Dissipative Phenomena in Condensed Matter**
Some Applications
By S. Dattagupta and S. Puri

72 **Predictive Simulation of Semiconductor Processing**
Status and Challenges
Editors: J. Dabrowski and E.R. Weber

73 **SiC Power Materials**
Devices and Applications
Editor: Z.C. Feng

74 **Plastic Deformation in Nanocrystalline Materials**
By M.Yu. Gutkin and I.A. Ovid'ko

75 **Wafer Bonding**
Applications and Technology
Editors: M. Alexe and U. Gösele

76 **Spirally Anisotropic Composites**
By G.E. Freger, V.N. Kestelman, and D.G. Freger

77 **Impurities Confined in Quantum Structures**
By P.O. Holtz and Q.X. Zhao

78 **Macromolecular Nanostructured Materials**
Editors: N. Ueyama and A. Harada

79 **Magnetism and Structure in Functional Materials**
Editors: A. Planes, L. Mañosa, and A. Saxena

80 **Micro- and Macro-Properties of Solids**
Thermal, Mechanical and Dielectric Properties
By D.B. Sirdeshmukh, L. Sirdeshmukh, and K.G. Subhadra

81 **Metallopolymer Nanocomposites**
By A.D. Pomogailo and V.N. Kestelman

82 **Plastics for Corrosion Inhibition**
By V.A. Goldade, L.S. Pinchuk, A.V. Makarevich and V.N. Kestelman

83 **Spectroscopic Properties of Rare Earths in Optical Materials**
Editors: G. Liu and B. Jacquier

84 **Hartree-Fock-Slater Method for Materials Science**
The DV-X Alpha Method for Design and Characterization of Materials
Editors: H. Adachi, T. Mukoyama, and J. Kawai

85 **Lifetime Spectroscopy**
A Method of Defect Characterization in Silicon for Photovoltaic Applications
By S. Rein

86 **Wide-Gap Chalcopyrites**
Editors: S. Siebentritt and U. Rau

87 **Micro- and Nanostructured Glasses**
By D. Hülsenberg and A. Harnisch

MIX
Papier aus verantwortungsvollen Quellen
Paper from responsible sources
FSC® C105338

If you have any concerns about our products,
you can contact us on
**ProductSafety@springernature.com**

In case Publisher is established outside the EU,
the EU authorized representative is:
**Springer Nature Customer Service Center GmbH**
**Europaplatz 3, 69115 Heidelberg, Germany**

Printed by Libri Plureos GmbH
in Hamburg, Germany